RESIDENTIAL BROADBAND

George Abe

Macmillan Technical Publishing
201 West 103rd Street
Indianapolis, IN 46290 USA

© 1997 by Macmillan Technical Publishing

Printed in the United States of America 2 3 4 5 6 7 8 9 0

Library of Congress Cataloging-in-Publication:
Available upon request

ISBN: 1-57870-020-5

Warning and Disclaimer

This book is designed to provide information about residential broadband. Every effort has been made to make this book as complete and as accurate as possible, but no warranty or fitness is implied.

The information is provided on an "as is" basis. The author and Macmillan Technical Publishing and Cisco Systems, Inc., shall have neither liability nor responsibility to any person or entity with respect to any loss or damages arising from the information contained in this book.

Editor-In-Chief	Jim LeValley
Executive Editor	Julie Fairweather
Cisco Systems Program Manager	H. Kim Lew
Marketing Manager	Kourtnaye Sturgeon
Managing Editor	Sarah Kearns
Director of Development	Kezia Endsley
Product Manager	Tracy Hughes
Development Editor	Laurie McGuire
Project Editor	Gina Brown
Copy Editors	Krista Hansing
	Molly Warnes
	Michelle Warren
Assistant Marketing Manager	Gretchen Schlesinger
Manufacturing Coordinator	Brook Farling
Cover Designer	Sandra Schroeder
Cover Production	Casey Price

Book Designer	Louisa Klucznik
Director of Production	Larry Klein
Production Team Supervisor	Laurie Casey
Graphics Image Specialists	Laura Robins
	Marvin Van Tiem
Production Analysts	Erich J. Richter
	Dan Harris
Production Team	Trina Brown
	Linda Knose
	Gina Rexrode
	Christy Wagner
Indexer	Kevin Fulcher

Trademark
Acknowledgments

About the Author

George Abe has been at Cisco Systems for three-and-a-half years and has been actively involved in Cisco's rollout of residential gateway and cable modems. Prior to his time at Cisco, George was at Infonet for 15 years and his last position was as the Director of New Technologies. George also has experience in sales, engineering, and product marketing.

Acknowledgments

Gratitude and appreciation to:

Laurie McGuire, Development Editor

Gina Brown, Production Editor

Michael Adams, Time Warner Cable

Mark Laubach, Com21

Mark Millet, Cisco Systems

Marty Schulman, Cisco Systems

Michael Colby, Ipsilon Networks

Ronald Holtmyer, Cisco Systems

Peter Ecclesine, Cisco Systems

Howard Mirowitz, Mitsubishi Electric

professionals all.

DEDICATIONS

To Rosie and Jimmy, who will be watching a different TV and surfing a different web than their Mom and Dad.

Contents at a Glance

Table of Contents

Introduction

On July 11, 1997, a startup company by the name @Home went public on Wall Street. @Home is a system integrator providing data services over cable networks. Stock was offered at $10.50 per share and rose to $25.00 during its first day of trading. By the end of the first few weeks, its market capitalization was established at over $2 billion. Not bad for a two year-old company with cumulative losses of $50 million and fewer than 10,000 customers. But such is the current interest, perhaps frenzy, of product development and investment surrounding a new product area called *residential broadband*.

What's going on here? Who, besides the backers of @Home, are cashing in and who is missing out? What are the issues? Residential broadband is a tremendously diverse subject, embracing high technology, government regulation in the public interest, and entertainment production values.

Residential broadband is the meeting ground for consumer electronics, the Internet, telecommunications, cable television, satellites, politics, and the film industry. How could there not be huge public and professional interest?

PURPOSE OF THIS BOOK

A basic definition is needed before this book proceeds. Residential broadband (RBB) networks are fast networks to the home; in particular, they are fast enough to provide some video service. This generally requires speeds of at least 2 Mbps.

Networks capable of such speeds are familiar to businesses and a lot is known about business networking. All large businesses seem to be wired today and many smaller ones are as well. There are many texts on local area networks, metropolitan area networks, telecommunications (for example, telephony), and the like. But as a much newer phenomenon, RBB networks have not yet received equal press. This book aims to provide a comprehensive, accessible introduction to residential broadband. More specifically, this book has the following goals:

- To describe new and existing entertainment and data services, and evaluate how demand for them will drive the development of RBB.

- To define basic technology requirements for implementing RBB and assess their state of readiness.

- To overview business conditions and regulatory practices that affect rollout and viability of RBB networks.

Although it contains some technical discussions, this is not a comprehensive engineering reference. The concern here is broader, and includes business and regulatory issues as well as technical ones. This is not a business networking book. RBB networking differs from business networking in several respects:

- **RBB emphasizes entertainment.** Residential users, unlike business users, require the delivery of entertainment.

- **RBB demands ease of use.** Residential users will have less access to professional support than business users.

- **RBB's scale is potentially huge in comparison to business networking.** The number of homes in the United States outnumbers the number of businesses (100 million households in the United States compared to about 10 million businesses and offices). Systems catering to residences must have superior scaling properties compared to business-oriented systems.

These characteristics of RBB are actually design goals, and will come up over and over again in this book. They are the ends that must be satisfied by technological, business, and regulatory means if RBB is to thrive.

This book is a survey of the most current thinking about RBB. It is not a crystal ball. The goal of this book is to provide a broad-based synthesis of available information—from technical documents, journals, popular press, direct involvement with product development and standards work—so that readers can make their own informed predictions about the future of RBB services. The scope of coverage here places some limits on depth, especially depth of technical detail, so another goal is to provide readers with sufficient direction and resources for their further exploration of the subject.

In stating both positive and negative issues affecting RBB, it is not my intention to promote or criticize any industry. In preparing this book, I have come to know and admire representatives of many companies and organizations involved in aspects of RBB. Throughout, I have been impressed with the effort, brainpower, and public spiritedness of these professionals in business and government, who will no doubt change the way we all receive and respond to information in our homes.

AUDIENCE

This book is aimed at anyone seeking a broad-based familiarity with the issues of residential broadband, including product developers, engineers, network designers, business people, persons in legal and regulatory positions, and industry analysts. Courses on residential broadband are beginning to be offered and this book may also be appropriate for students in an issues-based class. Hopefully, consumers—the people who will be buying and using RBB services—will find this book to be an interesting peek into what the future holds for them.

Because of its breadth, some information in this book may be familiar to some readers already. However, few readers are likely to be familiar with all the technical, business, and regulatory issues brought together here. The complexity and relative youth of RBB are such that anyone wanting to understand or guide its evolution should have some understanding of the full mix of influences on it.

In terms of background understanding that readers should have to get the most out of the book, this book errs on the side of giving plentiful background information instead of too little. Many engineering professionals in particular may find Chapter 2, "Technical Foundations of Residential Broadband," to be familiar information that they prefer to skip or only skim. It is included for the benefit of non-technical readers who need basic vocabulary as a foundation for the rest of the book. Throughout, chapter topics are clearly identified by section and subsection titles, so readers can opt out of reading any material with which they are already well acquainted.

ORGANIZATION AND TOPICAL COVERAGE

Chapter 1, "Market Drivers," overviews services that are driving the market for RBB, emphasizing digital broadcast television, Internet access, and the convergence of the two.

Chapter 2, "Technical Foundations of Residential Broadband," surveys technical issues—both challenges and solutions—that are shaping the evolution of RBB. It's especially intended to help non-technical readers get an overview of the technical issues without overwhelming them with details. This chapter presents a reference model on which specific network architectures in later chapters are based. The purpose of this chapter is to help readers recognize both the consistent and the distinct components and challenges from network to network.

Chapters 3–6 present overviews of different types of access networks, those services provided by telephone companies, cable network operators, satellite operators, and newcomers.

Chapter 7, "In-Home Networks," focuses on the residential portion of residential broadband—the technologies and challenges associated with the home network. Topics include inside wiring and new consumer electronics devices that are needed to deliver services provided by access networks to the den, the kitchen, the entertainment center, or wherever, in the home.

Chapter 8, "Evolving to RBB: Integrating and Optimizing Networks," covers the crossroads where access networks meet home networks. Software and system issues to be resolved in bringing these networks together are formidable. As a case study, the chapter looks at the Time Warner Pegasus service, now under development.

The Appendix, "RBB-Related Companies and Organizations," provides a list of equipment vendors, service providers, and industry groups—such as the standards organizations—including their web Universal Resource Locators (URLs) and stock ticker symbols, where applicable. The URLs point to what these organizations say about themselves. The stock tickers point to what Wall Street says about them.

FEATURES AND TEXT CONVENTIONS

Text design and content features used in this book are intended to make the complexities of RBB clearer and more accessible.

- Key terms are italicized the first time they are used and defined. Similarly, key abbreviations are italicized the first time they are used and spelled out.

- "At Issue" items discuss points of contention or ongoing industry debates. Where possible, they also include predictions about how contested aspects of RBB are likely to be resolved. Many readers are especially interested in such contentious issues but some are not. At Issue items may be skipped without sacrificing understanding of main text discussions.

- "Viewpoint" items are a venue for some of the more subjective aspects of RBB. Some of them represent my own opinions; in other cases they articulate the positions of other industry observers and analysts. Like the At Issue feature, Viewpoints may be read out of sequence from the main text at the reader's discretion.

- Chapter summaries, including summary tables, provide a snapshot of both problems and solutions that affect the future of RBB. Given the significant amount of information to be sifted through and the complex interplay between technical, business, and regulatory considerations, these summaries are meant to briefly reiterate the most pertinent issues.

- References to further information are included at the ends of the chapters. Although not all the references are cited directly in the chapter, all were useful to me in preparing this book. Those which are directly cited are bracketed in the body of the chapter and are listed in the references section.

TIMELINESS

During the course of preparing this book, numerous changes were required to reflect the most recent product rollouts, Supreme Court decisions, standards decisions, and technical innovations. In this rapidly evolving field, more changes will no doubt have occurred by the time this book reaches readers. Throughout, appropriate language has been used to indicate topics that were subject to change by the time of publication.

The more important solution to addressing the volatility of residential broadband has been to focus this book on important concepts that are not subject to rapid change. The interelatedness of technological, regulatory, and business issues—and the potential evolution of existing networks into residential broadband networks—are at the core of this book.

CHAPTER 1

Market Drivers

Consumers today receive a variety of services over multiple networks connected to the home. They receive voice and data services over telephone networks and broadcast television services through cable television networks and over-the-air. As technologies for the computer, entertainment, and communications industries converge into digital infrastructures, leaders in government and industry are contemplating the development of residential broadband (RBB) networks that will deliver these same services—as well as new ones—to the home.

Although RBB networks can offer great benefits to the consumer, these networks will be expensive to deploy. Service providers (telephone companies, broadcasters, and cable operators) and their equipment vendors are poised to make multibillion-dollar bets that consumer demand for broadband networks and the content they deliver will be high.

What will compel the consumer to purchase the new RBB networks? Consumers won't pay for RBB services just to have access to a fast network. Existing telephone and broadcast television services by themselves don't justify the rollout of high-speed residential networks; these networks already work reliably, are reasonably priced, and offer a comfort level for consumers.

To justify the rollout of residential broadband networks, new services or market drivers, must be available to the consumer over the new networks. *Direct broadcast satellite (DBS)* has shown that consumers will pay a premium for digital television reception. Digital television offers better pictures and a wider program selection than over-the-air broadcasters or cable operators can offer over analog networks. In addition, the Internet offers a tremendous variety of information and entertainment for the home. In fact, digital television and the Internet, offered separately or in combination, are creating new forms of entertainment and information services to the home that are among the market drivers for RBB networking.

This chapter will discuss several possible market drivers for RBB, including digital television, video on demand, near video on demand, the World Wide Web, and virtual channels. The following sections will focus on defining aspects of these services that are either facilitators of or challenges to deployment of RBB.

ANALOG TELEVISION

Digital broadcast television is viewed as a key market driver for RBB networking. But before considering digital television in more detail, it will be useful to review some important characteristics of analog television.

One can argue that analog television is the most prevalent communications medium in the world. The world has more televisions than telephones, and in the United States more homes have televisions than indoor plumbing. The following communication statistics provide you with a sense of how technology is affecting the worldwide population:

- Worldwide TV households in 1996: 909 million

- Worldwide PC households in 1997: 97 million

- Worldwide wired telephone households in 1996: 710 million

- Expected TV sales worldwide in 1997: 110 million

- Expected PC sales worldwide in 1997, including business sales: 90 million

- Television sets in the U.S.: 280 million

- Television sets in the U.S. per household: 2.9

- Telephones in the U.S. per household: 1.2

- Broadcast TV advertising revenues in the U.S.: $34 billion

- Total advertising revenue in the U.S.: $173 billion

- Viewing hours in the U.S. per year per household:

  - Broadcast stations   832 hours/year/household

  - Independent stations   196 hours/year/household

  - Basic cable   480 hours/year/household

  - Premium cable   87 hours/year/household

  - Total hours viewed in household   1,595 hours/year/household

  - Total hours viewed in household   4.37 hours/day/household

Not only globally, but also at the level of individual households, analog television is clearly a pervasive communication.

The Three Standards

Three standards currently exist by which to encode and transmit analog TV worldwide. The standards are the *National Television Standard Committee (NTSC)* (sometimes referred to as *Never Twice the Same Color*), *Phase Alternation Line (PAL)*, and *Séquentiel Couleur Avec Mémoire (Sequential Color with Memory or SECAM)*. Table 1–1 shows some major characteristics of each standard.

Table 1–1   *Characteristics of Analog Broadcast Standards*

	NTSC	SECAM	PAL
Total lines/screen	525	625	625
Active lines/screen	480	575	575
Pixels/line	640	580	580
Bandwidth/channel	6 MHz	8 MHz	8 MHz
Picture rate/second	29.97	25	25
Megabits/second (uncompressed)	221.2	400.2	400.2
Countries	USA Japan Canada	France French colonies Russia	Germany UK Rest of Europe

In the United States, TV channels are transmitted in increments of 6 MHz of bandwidth per channel. Channel 2 starts at 54 MHz, and Channel 6 ends at 88 MHz. (The 4 MHz of bandwidth between Channel 4 and Channel 5 are used for radio astronomy and radio navigation purposes.) From 88 MHz to 108 MHz, FM radio is transmitted. Above the FM band, TV channels resume at 174 MHz with Channel 7 and continue upward in increments of 6 MHz to Channel 13, which ends at 216 MHz. Channels 2–13 are the *very high frequency (VHF)* stations. In addition, viewers can receive *ultra high frequency (UHF)* stations. UHF stations begin at 470 MHz for Channel 14 and continue upward in increments of 6 MHz to Channel 69, which ends at 806 MHz.

Channelization rules are different elsewhere. In Europe, channel spacing is 8 MHz rather than 6 MHz, in either PAL or SECAM. In Australia, 7 MHz PAL encoding is used. More bandwidth per channel means clearer audio and video quality than with NTSC. Except for a given amount of aggregate bandwidth, fewer channels are available.

Analog Screens Versus Computer Monitors

One of the assumptions on which RBB relies is that content can be received on computer monitors as well as television screens. This assumption poses some difficulties because the entrenched standard, analog television, differs from computer monitors in two important aspects—display format and color coding.

Display Format

In all analog television, *interlacing* is used as the display format. Interlacing means that horizontal lines of pixels are illuminated in an alternating pattern rather than sequentially.

Figure 1–1 illustrates interlacing display. The television picture tube is a rectangular array of colored pixels that must be illuminated to present a picture. The dotted lines indicate the first illumination pass of the screen. Illumination moves from left to right at a slight downward angle. Then a line is skipped and the following line (the third line) is illuminated, the fourth is skipped, then the fifth is illuminated, and so on until the illumination reaches the bottom of the monitor. The entire vertical scan (called a *frame*) is performed 59.94 times per second.

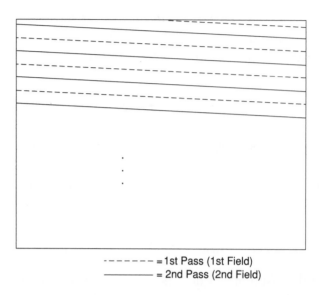

- - - - - - - = 1st Pass (1st Field)
————— = 2nd Pass (2nd Field)

The process starts again at the top of the screen with the other
lines being illuminated. The total effect is to refresh the entire picture 29.97 times per second. The fact that lines are skipped does
not affect the visual experience because the time spans and vertical spaces between the lines are too small for the human brain
and eyes to notice. The refresh frequency was chosen to reduce
the perception of flicker.

In *progressive scanning*, or *proscan* (used in computer monitors),
each line is illuminated sequentially without skipping lines. This
is a simpler approach than interlacing, and computer software
(such as graphics and fonts) has been optimized for proscan. For
computer monitors to display television signals, video cards must
perform *scan conversion*, which takes interlacing and renders it
on progressive computer monitors.

Proponents of interlacing—namely the broadcasters and television manufacturers—maintain that interlacing is superior to
proscan for two reasons. First, interlacing is a form of bandwidth

compression. Only half of the image is broadcast at an instant, therefore enabling the full picture to be sent in half the bandwidth. Secondly, proponents assert that interlacing offers better, softer pictures, especially for natural outdoor scenes. The human brain fills in the skipped lines, whereas progressive scanning shows too much, is too harsh, and is viewed as upsetting to the psycho-visual experience over long-term viewing.

Color Coding

Computer monitors differ from television in color coding. Whereas computer monitors use a red, green, blue (RGB) vector for each pixel to display color, television uses a luminance, hue, and intensity vector, called *Y'UV coding* for composite video and *Y'CrCb coding* for component video.

Television's system of color coding is an artifact of the transition from black-and-white to color. When television showed only black-and-white images, each individual pixel transmitted only luminance, or brightness. To make color television backward-compatible with black-and-white television, it was necessary to encode all colors by using a luminance vector. That is, instead of coding colors with varying amounts of red, green, and blue, it was necessary to code colors in varying amounts of luminance, as well as two other vectors called hue and intensity. The higher refresh rate for computer monitors is required because of higher contrast and luminance of display and wider viewing angle.

Peripheral vision is where the human eye is most sensitive to flicker. Table 1–2 summarizes the differences between televisions and computer monitors.

Table 1–2   *Color Television Versus Computer Monitors*

	Color TV Monitors Monitors (SVGA)	Computer
Scanning	Interlaced	Progressive
Color Coding	Luminance, Hue, Intensity (Y'UV)	Red, Green, Blue (RGB)
Pixels	640×480	800×600
Frames/second	29.97	72

Business Environment

Individual television stations are generally one of three types: network-owned and operated, affiliate, or independent. The network-owned and operated stations (O&Os) are owned directly by the big national networks—ABC, NBC, CBS, and Fox. Legislation exists, however, to prevent concentration of media power in too few hands. Until the Telecommunications Act of 1996 (Telecom 96), networks could own at most 12 stations, which were permitted to reach no more than 25 percent of the national television audience. With the passage of Telecom 96, no limits are placed on the number of stations, but each network's stations are permitted to reach no more than 35 percent of the U.S. population.

About 200 network affiliates exist for each national network. Affiliates are owned by independent business people and are therefore free to change their network affiliation from time to time. Affiliates are paid to accept network programming, which usually consists of prime-time programming, national news, weekend sports, and soap operas. The network sells advertising during the national programs, but affiliates can obtain their own programming, schedule permitting, and make much of their independently derived money on local news.

Finally, independent stations purchase programming from the networks and independent producers but are free to decide which specific programs they do or don't want, and when to broadcast them.

In total, more than 1,500 broadcasters exist in the United States. Each is represented by a trade association called the *National Association of Broadcasters (NAB)*, which functions as the industry's lobbying arm. You may visit the NAB web site at:

`http://www.nab.org`

Local content and advertising insertion are important services for affiliates and especially small independent stations that do not have a guaranteed flow of network programming. Local content is also important for the national networks because consumers like to watch local events, traffic, weather, and advertising. Local content and ad insertion raises new technical problems as stations make the transition to digital transmission. New techniques to merge national content with local content will be required, which consequently will require new investment.

The point of this discussion is that the transition from analog to digital transmission will be costly and that the majority of stations in the United States—namely the independents and some affiliates—will face financial hardship during the transition.

Regulation

In the United States, as elsewhere, the broadcast television industry is regulated by national authorities. In this country, the Federal Communications Commission (FCC) regulates many aspects of broadcast television. Among these aspects are ownership rules of stations, media control, spectrum allocation, and technical specifications.

In the past, the FCC and Congress have required backward-compatibility for TV innovations. The FCC, for example, mandated that color television sets be able to receive and display black-and-white signals when color television was just becoming available in the 1950s. In the 1960s, a congressional mandate required that TV sets be able to receive UHF as well as VHF stations over the objections of the television industry. Requiring the combined reception of UHF and VHF was viewed as excessive government meddling, but it created new channel capacity and opportunities for programmers.

Regulations in the public interest also exist with regard to Must Carry rules, foreign ownership of U.S. television stations, and closed-captioning. Must Carry rules require cable companies to carry local broadcast stations. In addition, foreigners cannot own majority shares of a broadcaster and cable operators are required to provide closed-captioning support. These rules indicate that similar regulations will carry over to digital television.

Key Pressures on Analog Television

Broadcasters, cable operators, television manufacturers, and the government are all interested in changing the existing analog television regime. Their various concerns can be summarized as follows:

- **Channel lineup and scarcity.** The Must Carry rules highlight a fundamental problem of analog TV, both over-the-air and cable. Insufficient channel capacity exists for all the networks that broadcasters and cable operators want or need to get on the air.

- **Security.** Cable operators experience significant pirating of signals, and analog scrambling has not been effective in stopping this security breach.

- **Picture quality.** Picture and audio quality could stand some improvement. Screens are getting larger, and the proliferation of channels is stimulating increased production of made-for-television film production. These programs should be presented more attractively.

- **Slow TV sales.** Even though roughly 110 million television sets are sold annually around the world, profit margins on television sales are poor. Relatively few innovations in recent years have motivated sales, particularly when consumers tend to keep TV sets for a decade or longer. The TV industry needs a boost.

- **Spectrum auctions.** From the government's point of view, analog TV is a wasteful use of spectrum. The government would like to see other, more efficient uses of spectrum emerge. Analog TV spectrum could be returned to the government and auctioned to feed the government's coffers in a relatively painless way.

For more complete discussions of analog TV, see [Jack] and [Watkinson].

DIGITAL TV

Originally, the U.S. development of digital TV was motivated by a desire to reclaim the television manufacturing industry from Japan and Korea. By providing better pictures—a new video experience called *High-Definition Television (HDTV)*—U.S. manufacturers hoped to re-establish their former market leadership in television production and sales. During the 10-year process of defining this new product, HDTV metamorphosized into a method to deliver other media to the consumer. HDTV can deliver data service and multiple channels where only one existed before. Therefore, digital television is an obvious market driver for RBB. Digital television also resolves many of the key challenges facing analog television, such as the aforementioned key pressures of analog television.

The next section focuses on the distribution of digital TV over-the-air, as defined by the FCC decision announced in December 1996 and approved by the FCC in April 1997. Other ways to distribute digital TV, such as cable TV, telephone networks, and direct broadcast satellite (DBS), will be discussed in Chapter 3, "Cable TV Networks," Chapter 4, "xDSL Access Networks," Chapter 5, "FTTx Access Networks," and Chapter 6, "Wireless Access Networks."

The Origins of Digital TV

By the middle of the 1980s, the United States pretty much had lost the television manufacturing market to Japan. Even with their competitive edge, Japanese manufacturers recognized the need to improve standard television to generate new products and to extend their market lead. One way to do this was to display better pictures, especially for large-screen televisions and large outdoor displays, such as those at baseball parks. Early in the 1980s, the government-operated Japanese national TV network, NHK, and their equipment suppliers embarked on a project to create HDTV.

The Japanese system, called *MUSE (Multiple SubNyquist Sampling Encoding)*, was an analog system using more than 6 MHz per channel. MUSE was therefore incompatible with existing television and did nothing to alleviate the channel scarcity problem. It did, however, have a nice picture quality.

In February 1987, the FCC formed the *Advisory Committee on Advanced Television Services (ACATS)* to provide input to the FCC via a competitive process on technology regarding advanced television. In the development process, HDTV changed from the Japanese analog model to a U.S. digital model, primarily due to pioneering work by Zenith Electronics, General Instruments, and the Massachusetts Institute of Technology.

The key facilitator in the shift to digital television was the development of digital compression techniques. Table 1–1 shows that standard NTSC televisions would need more than 220 Mbps of bandwidth to deliver pictures with existing pixel density and color; PAL and SECAM would need more than 400 Mbps. HDTV, compared to the MUSE system, would require more than a billion bps. These bit rates could not fit in a 6 MHz channel without compression.

The solution was found in the use of *Moving Picture Experts Group (MPEG)* compression. MPEG compression enabled HDTV to fit into 19.3 Mbps, a reduction in bit rate of more than 50 to 1, which could fit into a 6 MHz channel. This was a tremendous technical accomplishment, thought by many experts at the time to be impossible to achieve.

MPEG compression makes it possible to offer NTSC-quality pictures in less bandwidth than 6 MHz. Digital television with NTSC pixel resolution is known as *standard-definition television (SDTV)*. SDTV produces a slightly better picture than analog TV because of less ghosting and fewer vertical hold problems, and it uses only a quarter or a third of the bandwidth required by NTSC. By using less bandwidth, the broadcasters and cable operators could offer multiple channels of SDTV, or perhaps new services such as Internet access, on the 6 MHz they would use for a single analog TV channel or a single HDTV channel.

In November 1995, a partnership called the *HDTV Grand Alliance* presented its recommendation to the FCC. The members of this partnership consisted of representatives from AT&T, General Instruments Corporation, Massachusetts Institute of Technology, Philips Consumer Electronics, David Sarnoff Research Center, Thomson Consumer Electronics, and Zenith Electronics Corporation. The Grand Alliance recommended a modulation technique, video compression (both of which are discussed in Chapter 2, "Technical Foundations of Residential Broadband"), and display formats for HDTV. The recommendation was a digital technology that fits in a 6 MHz channel. (Japan, curiously enough, steadfastly refused to go digital for advanced television until well after the Grand Alliance agreement.)

On April 3, 1997, the FCC announced rules by which the nation's over-the-air broadcast television system would transition to digital TV. This ruling ended a process that had begun more than a decade earlier with the formation of the *Computer Industry Coalition on Advanced Television Service (CICATS)*.

The Computer Industry Responds

As previously noted, television screens and computer monitors differed with respect to display format and color coding. After the Grand Alliance presented its proposal to the FCC in November 1995, the computer industry—represented by CICATS, comprised of Apple, Compaq, Cray, Dell, Hewlett-Packard, Intel, Microsoft, Novell, Oracle, Silicon Graphics, and Tandem—weighed in with comments. This coalition wanted greater accommodation for progressive display and the removal of the requirement that the new digital monitors display HDTV. CICATS argued that a computer monitor should be able to receive a high-definition feed, but that its display—progressive or interlaced, high-definition or standard-definition—should be left to the designer and manufacturer's discretion. The FCC concurred and in December 1996—after nearly 10 years of discussion, development, and trials—announced the final accord acceptable to broadcasters, consumer electronics vendors, computer manufacturers, and the film industry.

Accounts of the entire evolution of HDTV are given in [Brinkley], [Wiley], and [Grand Alliance].

The Agreement

The agreement among broadcasters, consumer electronics vendors, the computer industry, and the film industry essentially provides that the 1995 Grand Alliance proposal be accepted, but a display format is not specified. The accepted Grand Alliance components are shown on the Advanced Television Service Committee web page:

```
http://www.atsc.org
```

The following list outlines the Grand Alliance components:

- **Video compression** is to be MPEG-2, with various compression options accepted. Multiple compression options permit service providers to exercise considerable control over bandwidth utilization and picture fidelity. (Chapter 2 discusses MPEG-2 in greater detail.)

- **Audio compression** is to be Dolby AC-3. This choice is somewhat odd because MPEG has an audio specification, but it was ignored.

- **Transport format** is MPEG-2 Transport Streams. This is the fixed packet-length option of 188 bytes, which is suitable for lossy networks over wide areas.

- **Modulation scheme** is 8-VSB. Using this technique, a 6 MHz terrestrial broadcasting channel can support a bit rate of 19.3 Mbps, which is suitable for high definition.

- **Video scanning formats** can be any of 18 accepted formats, which are summarized in Table 1–3 and Table 1–4.

Table 1–3 *Progressive Video Scanning Formats for Digital TV*

Vertical Lines	Horizontal Pixels	Aspect Ratio	Frame Rate per Second
1080	1920	16:9	24, 30
720	1280	16:9	24, 30, 60
480	704	16:9	24, 30, 60
480	704	4:3	24, 30, 60
480	640	4:3	24, 30, 60

Table 1–4 *Interlaced Video Scanning Formats for Digital TV*

Vertical Lines	Horizontal Pixels	Aspect Ratio	Frame Rate per Second
1080	1920	16:9	30
480	704	16:9	30
480	704	4:3	30
480	640	4:3	30

Two formats are emerging as leaders. These are 720 lines of progressive, called 720P, and 1080 lines of interlaced, called 1080I.

A broadcaster can transmit in any of the 18 accepted video scanning formats. The monitor receives all 18 formats but may display only a subset of them, depending on its design specifications. No monitor is required to display any particular format, including HDTV. So, for example, a digital TV can receive an HDTV feed but might display only SVGA format. In this particular example, the monitor must perform scan conversion and conversions on pixel density and aspect ratio. The digital TV is therefore a multisync monitor because a receiving monitor can be used for both television reception and computer display.

Because many computer monitors have 1280×1024 displays already, they have the pixel density to display high definition, whereas today's TV monitors cannot. Therefore, some computer advocates say that the first viewers of HDTV could use computer monitors rather than television sets as the viewing device. The 1280×720 Grand Alliance format, for example, can fit on a 1280×1024 computer monitor with room left over for picture-in-picture (PIP) enhancements or out-of-band text annotations.

The CICATS and Grand Alliance accord likely will benefit cable operators as well as over-the-air broadcasters because many of the components of a digital set top are common to both over-the-air and cable. A *digital set top* is a device that accepts digitally encoded television broadcasts and converts them to display on an analog TV. The radio frequency (RF) components would be the same, thereby reducing costs. The main differences between cable and over-the-air television are conditional access and modulation scheme. The exact configuration of a digital TV capable of over-the-air and cable reception is a work in process.

Initial Rollouts of Digital TV Service

At the 1997 CableLabs summer conference, CBS and HBO provided insight in their initial plans. CBS has chosen to transmit filmed material in 1080 progressive (24 frames/sec) and video material in 1080 interlace (60 fields/sec). 1080 progressive format and the 1080 interlace format (of the 18) HDTV formats have been chosen by the Asian Broadcasters Union, the European Broadcasters Union, and by the Japanese broadcasters to replace MUSE.

CBS is currently on the air for three hours a day in New York City and broadcasts from the top of the Empire State building. CBS has verified an 80-mile reception radius.

HBO plans to do the following:

- Transmit in the aspect ratio of the original material in HDTV. (So, for example, *Ben Hur* would be shown letter-boxed at its original 2.35:1 on a 16:9 HDTV.)

- Simulcast in true HDTV for all studio and HBO original movies; this makes up about 70 percent of HBO's programming. This will require conversion to HDTV of 850 movies per year. Currently, the film industry converts only 30 movies into HDTV format per year.

- Launch in third quarter 1998.

HBO has so far declined to specify their digital formats. It remains to be seen if their display formats will be compatible with the broadcasters, and, if not, what the cost implications on the monitors will be.

AT ISSUE

Standardizing Display Formats

Eighteen presentation formats exist for digital TV. It is possible for a television set to receive and display all of them, but this would be expensive. Therefore, the issue of what subset of 18 ATSC options will predominate must be resolved. The consumer electronics industry claims that a single chip, 18 format decoder will soon be available.

Some television set manufacturers will produce high-definition devices, whereas others will manufacture only standard-definition units. Still others will display progressive or interlaced units. For the sets to be reasonably and similarly priced, broadcasters, content providers, and manufacturers likely will need to come to some compromises on which formats they will present.

As illustrated in Figure 1–2, digital broadcasting will be comprised of several elements: a broadcast facility, an affiliate or cable system facility, and the home.

The broadcast facility is a national center where programming is originated. In the digital environment, content can be captured on digital cameras (which are yet to be marketed in quantity to the broadcasters) or derived from video servers, which store content digitally. Substantial analog production equipment (cameras, editors, and sound equipment) will be produced for a long time to come because analog content will need to be converted to digital format.

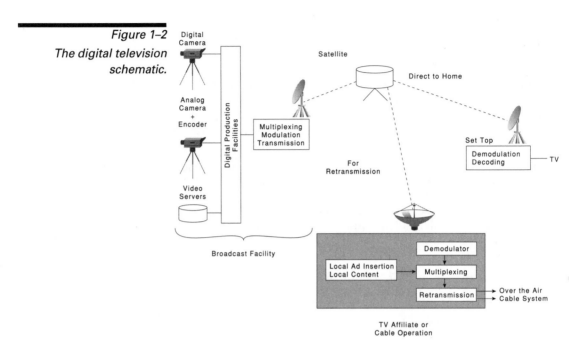

Figure 1–2
The digital television schematic.

New digital production facilities, such as those needed for editing equipment and tape machines, will be required to edit the digital content. After the content is produced, it is ready for transmission. In the digital environment, this means multiplexing multiple digital streams, modulating the streams, and transmitting them over satellite or microwave to local distribution.

Local distribution is provided by network affiliates or independent stations that are located near the viewer's home. The local station adds local advertising and programming of its own and merges its content with the national feed. The station then retransmits the program over-the-air or via cable. For some services, the digital content can be transmitted directly to the home via satellite without any intervening affiliates or cable systems.

The new components of digital TV systems are digital production equipment at the production end and decoders at the consumer end. Both ends require substantial investment, perhaps millions of dollars for the broadcaster and the rebroadcaster and a few hundred dollars for the consumer. The benefits of digital television, however, are expected to outweigh the costs.

Benefits of Digital TV

Although the issue of video display is not yet settled, the benefits of digital TV are clear. These include increased channel capacity, increased programming options, improved picture quality, heightened security, improved television sales, and potential auctioning of available bandwidth after redistribution.

Increased Channel Capacity

Analog TV occupies 6 MHz of bandwidth. SDTV channels can occupy as little as 1.5 MHz for equivalent viewing. A broadcaster can have three to five programs where he previously had only one. A cable operator or wireless cable operator with hundreds of MHz of bandwidth can offer hundreds of programs. In addition to providing more room for programming, increased channel capacity allows for increased advertising space and revenues, as well as new services such as Internet access.

Increased Programming Options

A corollary to increased channel capacity is new programming opportunities. In the future, for example, it will be possible to transmit multiple versions of an athletic event simultaneously. Multiple camera shots can be used by the telecaster, and the viewer can select the particular camera angle he prefers at the moment. American football frequently uses more than a dozen cameras to televise a game, and the director's job is to select a particular camera angle (blimp, huddle, wide angle, tight shot). With digital TV, it is possible for the consumer to make the selection. This could be especially useful with picture-in-picture televisions to grant consumers more viewing choices. Whether a charge will be levied for this enhancement or consumers will come to expect this service for free remains to be seen, however.

The Internet also has proven to be popular with residential users. An estimated 18 million residential customers subscribed to the web in the United States as of the end of 1996. If the web can be linked with digital TV on a single monitor, many possibilities exist for new forms of entertainment and information that emphasize the use of video. These new forms would increase viewership of digital television and help capture the legions of potential young viewers who are "surfing the Net."

One form of programming is to combine chat sessions with TV reception. As viewers watch a program, they could click a window on the TV and join a chat session with others watching the same show, or with production people who can provide insight into how the show was produced. This requires a two-way capability and presents some technical challenges. A television and chat session combination is attractive to programmers because the viewer provides content and marketing information.

Another form of content involves the time shifting of advertising. As viewers watch TV, they might see an advertisement they would like to keep and refer to later. In this case, the viewers could bookmark the commercial, similarly to how they bookmark web pages on a browser. Later, data could be transmitted from the browser to the content provider to activate the commercial and possibly add some other features, such as initiating a purchase order. The whole universe of order processing is a subject of great interest to advertisers and programmers.

Another form of digital programming is the virtual channel. A real TV channel occupies a certain amount of bandwidth and is transmitted to the consumer. A virtual TV channel, however, can be assembled on the customer's premises. A broadcaster can transmit computer commands that instruct a PC or a digital set top to create pictures. One can imagine a cartoon channel that transmits the equivalent of PostScript™ commands to the consumer. A consumer device accepts these computer commands and renders images and sounds. The *actual* images and sounds are not transmitted, but they are rendered in the home. This form of programming is useful for channels such as the Cartoon Network. In this form, the channel would occupy a fraction of the bandwidth required of a real TV channel.

Legions of creative talent in programming and advertising are experimenting with ways in which the web and TV can be combined. An interesting place to view this convergence is at the headquarters of Creative Artists, a major talent agency in Hollywood that has assembled a multimedia laboratory specifically designed to show their clients (writers, directors, and actors) the creative possibilities of the convergence.

At Issue

Free Versus Pay TV

What content and applications will be provided over free, over-the-air, advertiser-supported service? A key question is whether new broadcast digital services such as HDTV, multichannel SDTV, Internet access, and chat sessions siphon revenues from cable and satellite.

continues

Thirty or more channels of free television with some support for data services might be a strong competitor to pay television. Free television also might have options for picture-in-picture television and other advanced services. It remains to be seen what programming will be available and how popular it will be with consumers and advertisers.

Picture Quality

For most content, picture quality improves with digital transmission compared with analog transmission, even with standard-definition digital. This is because fewer problems occur with ghosting, vertical hold, and noise. Broadcasters, both over-the-air and cable, can encode with MPEG-2 at speeds from 3 Mbps to 9 Mbps for SDTV. With this amount of control, broadcasters have a means by which to improve picture quality incrementally as picture and audio content dictate. Chapter 2 discusses MPEG-2 compression in greater detail.

Security

Digitally encoded programs can take advantage of computer techniques such as *Digital Encryption Standard (DES)* encryption to scramble programming, thereby preventing unauthorized persons (people who don't pay) from receiving premium service. This problem is significant for both cable and over-the-air satellite services. Industry estimates put revenue loss due to piracy at more than $500 million annually. A theft-of-service sting operation for the Mike Tyson and Bruce Seldon heavyweight championship fight in 1996, for example, revealed an 18 percent piracy rate. Piracy rates such as this are particularly troublesome because boxing matches featuring Mike Tyson accounted for 50 percent of all pay-per-view revenues generated by cable operators in 1996.

Improved TV Sales

With digital TV, consumer electronics manufacturers have the incentive they need to increase TV sales. There are about 280 million analog TV sets in American households; with better picture and new features, digital TV provides an incentive for viewers to replace their analog sets. In addition, digital TVs will be able to connect to personal computers, thereby expanding the TV market by competing with computer monitor vendors.

Spectrum Auctions and Local Government Use

Finally, in the United States, analog spectrum currently occupied by broadcasters is valued for its auction potential. Portions of the 402 MHz of prime real estate (the VHF and UHF frequencies) will be auctioned with proceeds going to federal coffers. Other portions of the analog spectrum will be provided to local authorities for police and fire use. The sooner the analog frequencies are made available by the emergence of digital TV, the sooner the federal budget improves and local governments get more bandwidth. This has Congress particularly anxious to get on with the conversion to digital TV.

Transition to Digital Over-the-Air Television

The announcement made by the FCC on April 3, 1997, was truly historic for American television. This resolution sets a specific timetable for the transition of all TV to digital and lays out regulations for spectrum usage. This section overviews these transition issues and includes an example of how they will play out in one market.

Timetable

The major dates for the transition from analog to digital television are as follows:

- By Christmas 1998, the top-10 television markets will have digital broadcasts from the major networks. Viewers in these markets will have at least three digital channels from which to choose.

- By 2001, all commercial networks will broadcast in digital.

- By 2002, all public broadcasting networks will broadcast in digital.

- By 2006, analog spectrum will be returned to the FCC for auction.

New television sets will appear on the market in late 1998. In 2006, analog television goes blank. Consumers who have analog TVs at that time will need to buy set-top converters, as they do for cable today.

Figure 1–3 summarizes key dates and events in the history of analog and digital television.

Figure 1–3

Time lines for analog and digital television.

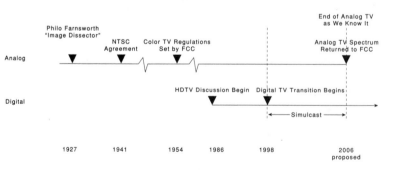

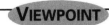

Goal Date of 2006

Despite the language of the Telecommunications Act of 1996, the date of 2006 to terminate analog transmission certainly will be flexible. Given the more than 280 million analog sets in place now, if 10 million digital sets a year are purchased, then less than a third of all sets will be digital-enabled by 2006. According to the Balanced Budget Act of 1997, if more than 15 percent of the viewing public does not have digital TV by 2006, Congress is authorized to extend the transition date. Depending on the rate at which consumers purchase digital TVs, it could take many years beyond the appointed date before analog transmission can cease. Programming incentives are needed for viewers to buy digital equipment to keep the process on schedule. This is why many observers feel that HDTV is necessary; it could be the best incentive to persuade consumers to get a new TV set.

Spectrum Usage

The Telecommunications Act of 1996 stipulates that broadcasters will not pay for the digital spectrum. Instead, the spectrum will be distributed free rather than auctioned. In return, broadcasters must provide free, over-the-air, advertiser-supported service. It is Congress's intent that there be free over-the-air digital TV, and free digital spectrum is an expression of that intent. If the broadcasters unwisely decide to encrypt channels and charge for subscription service on the digital spectrum, they will be forced to go to competitive auction for that spectrum.

Also in return for free spectrum, new public service requirements will be established, such as provisions to provide free political campaign advertising on the digital channel as part of a campaign reform package. Details of this are to be determined in Congressional proceedings on political campaign reform.

The broadcasters are free to use the digital spectrum in any manner they choose, which is to say that SDTV is permissible, thereby providing the broadcasters with a multichannel capability. Until the year 2006, when the analog spectrum is returned to the government, all the nation's broadcasters will simulcast their content in analog and digital formats, beginning whenever they can get digital transmission equipment installed. They will continue to broadcast in today's analog to keep consumers with analog sets viewing, but they will simulcast in digital for those with digital sets. HDTV and program multiplexes, which are market drivers for RBB, will be available only in digital form.

Transition to Digital in Los Angeles

Table 1–5 shows the channel assignments for digital television in the Los Angeles area as listed in the FCC's Order, Advanced Television Systems and Their Impact upon the Existing Television Broadcast Service, Mass Media Docket No. 87-268, released April 21, 1997 (FCC 97-115) and located at:

```
http://www.fcc.gov/oet/headline/fcc97115/Welcome.html
```

Table 1–5   *Digital Channels for Los Angeles*

Analog Station	Digital Station	Power (kW)	Area (sq km)	Population	Coverage (percent)
2	60	828.	39943	13,460,000	81.1
4	36	680.	41063	13,830,000	84.3
5	68	1000.	38228	13,519,000	80.8
7	8	10.	34851	13,722,000	95.5
9	43	342.	23622	12,774,000	94.6
11	65	659.	32990	13,278,000	94.1
13	66	650.	32263	13,186,000	95.0
22	42	165.	16523	11,629,000	92.1
28	59	182.	25452	12,719,000	99.6
34	35	70.	22216	12,586,000	99.7
58	41	55.	21665	12,534,000	100.0

KCBS Channel 2, for example, is the CBS affiliate in Los Angeles. This station is now on Channel 2 and will be simulcast on Channel 60 for digital reception. Channel 60 will transmit at 828kW. Because of the leap in frequency utilization from Channel 2 (54 MHz) to Channel 60 (752 MHz), Channel 60 will transmit at roughly 25 times the power of Channel 2 to achieve comparable geographic coverage. This significantly increases the station's electric bill. Even so, only 81.1 percent of existing Channel 2 viewers will expect to receive comparable reception of Channel 60.

CBS management, of course, will make certain modifications to ensure no loss of coverage (such as antenna shaping, retransmission from other CBS affiliates, cable, and microwave). But this example illustrates the technical and cost issues facing broadcasters when they simulcast analog and digital. Significant costs will be paid by the station without any incremental increase in audience. Each station will have its own unique financial circumstances to consider.

At Issue

Spectrum Flexibility

When broadcasters face the issue of programming for their new digital channel, a basic question is whether to broadcast in high-definition or standard-definition. If they opt for standard-definition, they then must face the questions of which collection of programs will be offered and where the content will come from.

Many stations have access to filmed content, particularly stations partially owned by film studios such as Turner Broadcasting. Filmed content is particularly amenable to HDTV because it is already in progressive format and HD aspect ratio.

Stations that don't have access to filmed archives will have to purchase or create more original content, or more likely, go to syndicated, home shopping, or adult programming. This content is more likely to be in standard-definition format because it does not need high-definition resolution and can be produced cheaply.

continues

With SDTV, each analog channel translates into several free digital channels over-the-air. Content-rich broadcasters might opt to go standard to have more venues for their programs. Disney (which owns Capital Cities and ABC), for example, could elect to transmit ABC, ESPN, and the Disney Channel over-the-air on spectrum that previously afforded only one channel. This is called *multiplex programming*. With multiplex programming, over-the-air broadcasters can compete more closely on the basis of channel capacity with cable and satellite operators, perhaps even carrying content that is currently available only on those outlets. Cable and satellite operators, for example, presently are the only outlets for ESPN and the Disney Channel.

National networks might find themselves in conflict with affiliates if they use multiplex programming, though. If, for example, Disney decides to use multiplex programming by broadcasting ESPN over-the-air, this could have negative consequences for the ABC affiliate ratings.

At issue, therefore, is the range of programming choices open to the broadcasters and how these choices affect the emergence of RBB.

Convergence of Internet and Digital Television

Computer issues might have been taken into account late in the process of the Grand Alliance's planning, but today these concerns are omnipresent. In particular, the Internet's tremendous growth already is influencing digital TV.

The convergence of the Internet and digital TV takes two forms. First, some cannibalization of TV viewership by the Internet does take place. Young viewers spend more time on the World Wide Web and less time watching TV. If the broadcasters, cable operators, and their advertisers want to continue reaching these people as extensively and effectively as they have in the past, they must do so at least partly via the web.

Second, technical convergence also exists. Personal computers and digital TVs will be increasingly difficult to differentiate. Modern PCs have video receiver cards, and TVs in the near future will have embedded microprocessors. These embedded microprocessors will be required for *electronic program guides*

(EPG) and digital decoding, so only a modest incremental cost will be required to add a little more horsepower and memory, and thus make a real computer out of a TV.

New services such as WebTV™, recently purchased by Microsoft, and Intercast™ already are blurring the distinction between the web and TV. WebTV offers Internet access by using an existing analog TV set without the need for a computer. Intercast offers TV viewing on a personal computer, with integration of web access. For more information regarding WebTV and Intercast visit their respective web sites at:

```
http://www.webtv.com
http://www.intercast.com
```

Traditional broadcasters have learned that linking web experience to television can expand web usage. In the 1996 Super Bowl, for instance, NBC had a promotion on TV to access a Super Bowl web site at half-time. More than three million hits were recorded. A similar promotion during the 1996 Summer Olympics yielded 14 million hits per day. If the web and television can be linked consistently, broadcasters can create a new form of entertainment.

Key Challenges for Digital Television

From the perspective of broadcasters, manufacturers, and content creators, developing digital TV brings both challenges to overcome and benefits to reap. Until the interested parties determine how to optimize their cost and benefit scenarios, the form and timing of digital TV remains unclear.

More than 1,500 broadcasters operate in the United States, and the majority of these are not affiliated with the major networks. Without the financial resources of a major network behind them, many of the smaller broadcasters will be challenged to deal with the financial realities of conversion.

The first investment for broadcasters is a new tower and transmitter. Digital TV production equipment—such as encoders, cameras, monitors, editing equipment, transmitters, and the like—could cost up to five million per station. As noted earlier, the electric bill for many stations also will increase substantially. The process of beginning simulcasts is likely to be quite staggered for purely financial reasons.

In some cases, the cost of going digital exceeds the book value of the station. In other words, some broadcasters will find it financially infeasible to go digital. Their broadcast licenses might well be sold to entities not currently associated with broadcasting, such as special-interest groups (including religious and political organizations) or corporations (such as the real Microsoft Network).

To complicate matters, digital TV sets will be more costly than analog models in the foreseeable future. Broadcasters must decide how to get consumers to pay a $500 or greater premium for a digital multisync TV. Either HDTV or new programming options would seem to be necessary. Further, a lack of digitally encoded content exists. Much work must be done to get the world's film library recorded to digital form.

Significant work also remains to facilitate interoperability of digital television. Decoders for DBS, cable, and over-the-air will use incompatible modulation techniques and security arrangements. It remains to be seen what this does to cost and customer satisfaction.

Of course, the fact that much of the push for digital TV does not come from consumers cannot be ignored. The movement to digital TV instead comes from television manufacturers who want to market a new generation of consumer electronics, from governments who want to derive revenues from spectrum auctions, and from programmers who seek new distribution for their shows.

The process of going digital is not new. Even with its incentive to go digital to compete with DBS, the cable industry has found the transition painful. Product development, deployment of customer units, costs to the consumer, content conversion to digital, security standards, and coexistence with analog service all have been nettlesome problems that the broadcasters have noted.

VIDEO ON DEMAND

Digital television has been discussed in a fair amount of detail because it is one of the most talked-about, most significant market drivers for residential broadband. Several other emerging services, however, also might prove to be important in fostering a market for RBB. One of these is *video on demand (VoD)*, a service currently under trial.

Service Description

VoD enables consumers to order movies over a network rather than going to a video rental store. The sale and rental of video cassette copies of old and new motion pictures, called home video, is a $16.2 billion per year market in the United States, much greater than the $5.5 billion per year for domestic box office. Already, home video produces about 12–15 percent of the total revenue of a given film industry release. The scale of this market lures broadcasters and cable operators to consider VoD as a substitute for home video.

VoD service is a *pull mode* service. Pull mode refers to the delivery method in which the subscriber demands and receives data from the provider. The consumer decides what to watch and when to watch it from a range of alternatives and then retrieves the selection. This is much like pulling information from a database.

VoD includes VCR controls (sometimes called *trick mode*), such as rewind, pause, and fast forward. VoD also has options for jumping to selected scenes, choosing language and subtitles, and activating captioning. It is widely believed that for VoD to be successful, features beyond simple VCR controls are necessary.

VoD's pull mode delivery stands in contrast to *pay-per-view (PPV)*, which operates in *push mode*. Push mode refers to the data delivery method in which the service provider transmits data to the subscriber on a fixed, predetermined schedule, or in response to some event such as the updating of data in the provider's database. The consumer simply decides whether or not to partake. Figure 1–4 depicts the conceptual differences between pull mode and push mode.

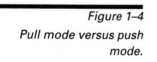

Figure 1–4
Pull mode versus push mode.

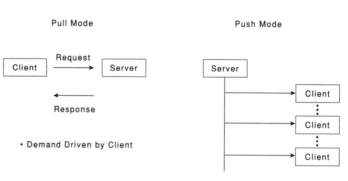

In push mode, no VCR type controls are available. Given the service characteristics of push mode, it is especially appropriate for live, one-time events such as boxing and wrestling matches, which are the two most popular forms of PPV in the United States.

In a pull mode environment, no two subscribers are likely to be watching the same movie or using the same trick modes at any given moment. Therefore, separate data flows are established, one for each viewer. Each data flow consumes bandwidth dedicated to a single consumer.

VoD was made feasible by the development of digital compression. Without compression, a single-color movie would consume perhaps as much as 100 GB of storage and could not be transmitted, even over a T3/E3 link to the home. With MPEG, a movie can be stored in 3 GB (providing broadcast TV quality) and played out at 3 Mbps. These are still large numbers, but they're double over proposed networks.

The measure of the success of a VoD offering is the *take rate*. The take rate is the number of movies rented per month divided by the subscriber base. If a subscriber base has 1,000 consumers, and 2,000 movies were rented in a month, the take rate is 200 percent. Take rates of 200 percent or more are necessary to make VoD financially viable.

Key Benefits of VoD

VoD is a convenient and highly customized way to view stored content such as movies, documentaries, and other educational fare. Not only does VoD give the viewer access to larger libraries than are available at a single retailer, but it also is a convenient means of access because it provides search mechanisms to locate specific topics. Furthermore, VoD is highly customized because it enables the viewer to choose what and when to watch instead of having the service provider decide.

Storage prices are dropping and access is improving. The cost of the technology is even improving, whereas the cost for video rental and sale are unlikely to drop further.

Finally, the content of VoD service is not speculative or risky, based on the proven success of the market for video rentals and sales.

Key Challenges to VoD

Royalty payments for content make up about half of the cost associated with VoD. The other half is engineering costs. Providers must have terabytes of stored data to have a marketable library, and the network must have good bandwidth-reservation characteristics. The essential characteristic of pull-mode delivery is that every viewer has his own connection to the service, which is not shared with any other viewer. Therefore, each viewer has his own network connection and bandwidth. If one person is watching a movie, the network must support, say 3 Mbps. If 10 people are watching, 30 Mbps is required. Each connection is expensive, and the lack of sharing creates relatively few economies of scale for the carrier.

A robust conditional access system is needed to prevent unauthorized viewers from watching programs for which they didn't pay. There has been, however, a cottage industry for years of people breaking cable security systems. More resilient procedures are needed as the economic stakes of VoD grow.

Viewer preferences pose another challenge to VoD. Viewers like watching a rented movie more than once or want to watch only part of a movie, preferences that cannot be accommodated economically by VoD. One partial solution is to permit the viewing of a single movie over an extended period of time. A two-hour movie, for example, may be viewed over a three-hour time period. This provides enough time for the viewer to have frequent rewinds.

As with other digital distribution methods, VoD is hampered at the moment by a lack of available content. Existing VoD libraries are not extensive enough; only a small fraction of the 16,000 movies in film vaults have been digitally encoded. Encoding pictures is a time-consuming and expensive process.

Experience from hotel movie rentals and various consumer trials indicate that, given a sufficiently large library of content to choose from, VoD can be a popular consumer service. It is expensive to deliver, however, and high costs have service providers looking for alternatives.

NEAR VIDEO ON DEMAND (nVoD)

An alternative to VoD is *near video on demand (nVoD)*, also known as advanced or enhanced pay-per-view or staggercast. Near video on demand provides widespread availability of movies without the need for a dedicated point-to-point connection between the viewer and the video server. This reduces server and network resources when compared with VoD but at a loss of consumer flexibility. Given experience with PPV, consumers appear to be willing to accept these limitations. With nVoD, the viewer is offered the same content continuously, whereas for DBS or PPV, the content is offered once. As such, nVoD is an experimental service undergoing consumer research.

Service Description

With nVoD, the service provider elects to offer a particular movie beginning at certain intervals, say every 15 minutes, on a small number of channels, such as four. The interval is called the stagger time, hence the term *staggercast*.

The network provider decides when to initiate the movie and on which channels. The subscriber then tunes to one of the four channels. If the viewer watches the movie from beginning to end, this viewing experience is exactly the same as tuning to a movie on any broadcast channel.

If the viewer wants the equivalent of rewind, however, she can switch to another channel that started 15 minutes later. If she wants only five minutes of rewind, though, she's out of luck. For fast forward, the viewer can tune to another channel that started 15 minutes ahead of the program stream she first started watching. This method does not offer precise rewind or fast forward, but the objective is to reduce cost by reducing the number of dedicated user connections, each of which consumes identical bandwidth. Four channels can service an entire viewing area using nVoD, because each of the four channels operates in broadcast mode. This is in contrast to VoD, where four channels can service exactly four viewers.

Stagger time can be reduced to five or 10 minutes but this increases the number of channels allocated per movie.

Figure 1–5 depicts the operation of nVoD. In this example *Gone With The Wind* (GWTW) is offered with 15-minute staggers over four channels.

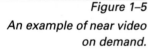

Figure 1–5
An example of near video on demand.

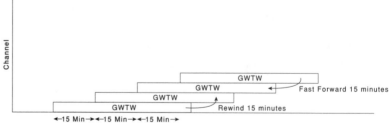

The nVoD technology fits the definition of push mode. The provider pushes the bit stream to the subscriber rather than the subscriber pulling bits from the server for exclusive use, as is the case for VoD. The royalty, compression, and set-top costs are the same as VoD, but substantial server and network savings result.

A typical stagger time between showings of a movie is 15 minutes. Because movies last about two hours in length, eight channels are needed for an even distribution of 15-minute staggers. The viewer has a limited amount of time to watch the complete movie, say, six hours. This permits a lot of rewinds. Service providers typically offer three different nVoD movies at a time, thereby requiring 24 digital channels for the total nVoD service.

Key Benefits

Compared to VoD, nVoD offers substantial savings to the service provider by reducing server and network resources. Eight channels provide service to exactly eight viewers by using VoD. But eight channels of nVoD can service a viewing population of thousands with 15-minute staggers. This is especially useful with recent releases, where a high probability exists that many viewers will be watching the same thing at roughly the same time.

Key Challenges to nVoD

Despite its benefits, though, nVoD still requires a large number of channels. In our example, 24 channels were needed to show only three movies. Cable and DBS operators would need to allocate channels to VoD or nVoD—channels that otherwise would be used for standard broadcasting. Because many operators already are strained for channel capacity, will they allocate enough broadcast channels to support an nVoD service? Or will they prefer to offer more network programming instead?

At the consumer end of the service, will consumers accept the loss of control to which they have grown accustomed with their channel changers? Fifteen-minute staggers might not satisfy users who want finer granularity of rewind and fast-forward timing.

As with VoD, nVoD must pay royalties and deal with a lack of digital video content.

THE WORLD WIDE WEB

In addition to video services, the web is a major market driver for RBB networking. As of December 1996, an estimated 18 million residential users in the United States use the web for work at home, educational, and recreational purposes. As web content becomes more tailored to fast networks—for example, by including more video—some practical obstacles to further scaling the web to consumer market proportions must be considered.

Web Browsers as Broadband GUI

Web browsers are already firmly entrenched as universal graphical user interfaces (GUIs) for narrowband services such as e-mail, file transfers, transaction processing, and web surfing. Table 1–6 projects the consumer Internet market for the near term in the United States.

Table 1–6   *Consumer Internet Market*

	United States households	Computer households	Internet households
1997	99.9	42.0	25.1
1998	100.9	45.0	33.5
1999	102.0	48.0	40.7
2000	103.1	50.0	47.8

Source: Veronis, Suhler and Associates

Browsers easily can be extended to become broadband GUIs. A browser, for example, can be used to make a phone call, without the caller's needing to know the destination phone number. All the caller needs is the universal resource locator (URL) for the business, and a button on the screen interface initiates the phone call. Callers usually find it easier to remember the URL than a phone number.

Similar interfaces can be performed for electronic program guides (EPGs), VoD content orders, and controls for VoD such as rewind, pause, and fast forward.

The web uses a transport protocol called the *Hypertext Transfer Protocol (HTTP)*. With HTTP, a user clicks an icon or an active link, and the browser establishes a connection to a database referenced by the icon or active link. The pointer to the referenced data is held at the server, which means that the client doesn't have to know anything about its location. The relatively seamless connection setup (which insulates the client from knowing the whereabouts of information) and the existence of search engines (which help users find data) have made the web easy to use, powerful, and the universally accepted method of accessing databases on the Internet.

Key Issues of the Web and Suggestions

Although millions of users browse the Internet, concerns exist as to its scalability to substantially larger audiences. Scalability problems will reduce the web's role as a market driver for RBB. Among the concerns are congested servers, long holding times, and high signaling rates. An approach to server congestion known as web caching is discussed in Chapter 8, "Evolving to RBB: Integrating and Optimizing Networks."

Holding Times

Holding time is the duration of a connection, during which resources, such as buffers, bandwidth, and control variables (such as indexes into databases) are held for the exclusive use of the client. Long holding times negatively affect the number of users who can be serviced simultaneously by the end-to-end system.

Web sessions tend to be longer in duration than voice sessions. This creates problems for telephone companies, which provide transit for data calls from modems or ISDN connections. These calls go through voice telephony switches. Call lengths have increased from about three minutes per voice call to more than 20 minutes per data call. Longer holding times through the voice switch increase the blocking probability per call (measured in probability of a busy signal) and eventually force the telephone company to buy additional voice switches to maintain grades of service mandated by their respective *public utility commissions (PUC)*. Voice switches are much more expensive than data switches, so telephone companies have a strong incentive to off-load the data calls from voice switches to data switches or routers.

The solution is to create access networks, such as cable TV (see Chapter 3) or the new generation of digital subscriber loop technologies (see Chapter 4), neither of which need the services of voice switches for data traffic.

Signaling Rates

Signaling refers to mechanisms by which one entity, such as a user application, communicates with another entity, such as a network, for control purposes. Examples of signaling are connection setup, (such as picking up a phone to make a voice call), connection release (such hanging up the phone in the voice scenario), or requests for some service characteristic (such as resource reservation).

Signaling is difficult and slow. Signaling rates, such as connection setups per second, are measured in dozens per second. State-of-the-art ATM switches, for example, can accomplish no more than a hundred or so call setups per second. These numbers compare unfavorably with a router, which has no call setup process for many applications and thus can move hundreds of thousands of packets per second.

Web access imposes strong signaling requirements on networking systems. Some experts predict a signaling crisis in networking systems because of the inability of networks of all types to support signaling events calibrated in the hundreds or thousands per second, the scale required of RBB systems. Mechanisms must be found to reduce signaling rates for RBB to scale.

PUSH MODE DATA

One way out of the signaling problems evidenced by the web is to move from pull mode to push mode data. Of course pull mode will continue to grow because it has advantages over push mode for ease of use and power. For many applications, however, going to push mode provides an opportunity for scaling to RBB proportions and providing important new revenue sources for the web.

Push mode data is to HTTP what nVoD is to VoD. Push mode is a bandwidth-reduction and signaling-reduction technique that permits greater scaling of the servers and networks that provide content. The tradeoff in both cases (push mode data and nVoD) is reduced user control and flexibility. For many applications, such as stock quotes, news, and software updates, the tradeoff for push mode makes sense.

An Example Using Software Updates

One potential application of push mode data is software updates. Many popular software programs, such as web browsers, are retrieved from web servers. To extract the latest version of Netscape browsers, for example, the user goes to the Netscape site and asks for the update. The user then is presented with a list of servers, from which the user selects the source of the update. Because of the popularity of Netscape browsers, the company needs to have multiple servers distributed worldwide; the user picks one, which presumably is closest in proximity. The user chooses a site, hoping that it is not too busy at that time. (This is a kind of user-initiated load balancing.) Having done so, a connection is set up to that server, and the update is sent.

This process imposes a number of problems. First, hundreds of requesters might converge on a single server at once. When you pick your server, you do not know the usage on that site. Perhaps a geographically more distant server has less congestion and therefore can offer better service.

More importantly, to retrieve your update, a connection is established with the server, and a dedicated bit stream is created between you and the server. If your colleague in the same building is retrieving the same update at the same time from the same server, you two will be duplicating use of server and network resources unnecessarily.

Push mode offers a way to reduce the duplication. Using a technique in push mode called *IP multicast*, staggered streams of data from a single server can be sent on regular intervals similarly to nVoD. Netscape could send software updates on a handful of multicast groups at 15-minute intervals, the same way in which a broadcaster sends a movie on 15-minute staggers. The user who wants an update joins a multicast group that begins shortly after he has notified the server. The data flow might not start immediately, but it would start in at most 15 minutes. The total load on the Internet would be substantially reduced, and only a few servers would be required instead of the dozen or so that Netscape currently has in North America alone. IP multicast is discussed further in Chapter 2.

For now, suffice it to say that multicast reduces bandwidth requirements through the network because the minimum number of packets are replicated to service a multiplicity of receivers. Mulitcast also reduces connection requirements, which alleviates overhead imposed on the network and the server.

Models for Push Mode Data

PointCast is a popular screen saver product that provides a data-casting service. *Datacasting* is the one-way transmission of data from one source to multiple receivers, similar to television broad-casting. Datacasting is achieved by using IP multicast over wired networks or by transmitting data over-the-air. PointCast gives the look and feel of push mode products and provides a basis for imagining how they might evolve. Visit the PointCast web site at:

```
http://www.pointcast.com
```

PointCast enables each viewer to receive a customized, point-to-point stream. The service is in push mode in that the server sends requested data without an explicit command from the client on a connection basis. The server transmits on an event basis by using a point-to-point connection. As a result of Point-Cast's point-to-point nature, lots of data is generated through the network. (Some companies have put controls on PointCast's use on their corporate networks because of the amount of data it generates.)

Castanet, a product from Marimba (`http://www.marimba.com`), takes the push mode concept a step further by enabling the user to identify in advance the servers from which he wants to receive data updates. The user also can specify time intervals at which to receive data that has been updated on the server. When data is sent, it is through a point-to-point connection from a networking perspective.

PointCast and Castanet do not take full advantage of the benefits of push mode because they do not utilize a point-to-multipoint (multicast) approach. Both products, however, serve as models for a true push mode data service.

Variations of Push Mode Services

Opportunities exist for a variety of push mode products. Delivery of stock quotations, for example, comprises a natural push mode service because all receivers would be interested in the same information. It, however, would be unfortunate if, owing to a networking problem such as congestion or a link outage, some receivers were unable to receive their quotes. It would be impractical for each receiver to request retransmission from the sender since there may be many such receivers. To avoid this problem, an extension to push mode is needed so that each receiver can reconstruct the damaged packet without requesting retransmission. Such techniques use forward-error correction techniques, which is discussed in Chapter 2. Such a system is known as *reliable multicast*, the development of which is currently in progress.

Another extension would be for receivers to synchronize reception. If, for example, stock quotes are coming from New York and receivers are in New Jersey and Tokyo, both receivers should get the service within a relatively short time of each other. This can be referred to as synchronized multicast.

An interesting problem related to multicast is determining what happens when the receivers do not have uniform characteristics. Some receivers, for example, might have 28.8 Kbps modem access to a flow, while others might have a leased T1/E1 line. How are they both to be accommodated if there is a common feed to which both will subscribe? One technique lies in decomposing the highest-resolution bit stream into multiple slower bit streams. The high-resolution service can be considered as the sum of several feeds, each of a cruder rendering. Each receiver can accept one or more of the cruder feeds according to how much bandwidth it can access. The receiver then binds packets from the member feeds to render a higher-resolution display. This is depicted in Figure 1–6 and can be referred to as the problem of heterogeneous receivers.

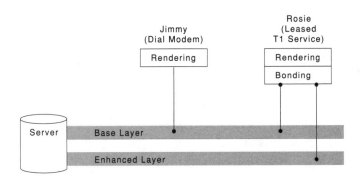

Figure 1–6
Heterogeneous receivers of push mode.

This example shows a content stream consisting of two bit streams and two receivers. Jimmy has the dial modem and Rosie has the leased line. The server emits a full-color MPEG-encoded video rendering. The Base Layer feed has a basic level of encoding of the bit stream and is transmitted at 28.8 Kbps. Jimmy receives that feed and gets what he can get, namely, the lowest common denominator rendering. Rosie receives two feeds. Like Jimmy, she receives the Base Layer, but she also receives an Enhanced Layer, which adds more bits. She bonds data from the two feeds to get high resolution.

For readers familiar with IP multicast, each of the layers corresponds to a multicast group. In this case, Jimmy joins one group and Rosie joins both groups. Work is in process from a company called VXtreme (`http://www.vxtreme.com`) to perform services such as this for IP multicast.

Benefits of Push Mode

Because the server controls the transport in push mode, this technology can be designed to emit long packets and use longer data flows. Longer packets provide more efficient bandwidth use, and longer flows reduce signaling overhead. The subscriber signals which multicast group it wants to join (digital tuning), but after that no requirement exists by which the subscriber initiates a call setup process, thus minimizing signaling.

Because the service provider controls the content and timing, push mode is more compatible with the advertising model used in television than is pull mode. It is possible that push mode will attract the advertisers and financing needed to make digital video and other high-cost services on the Internet profitable.

MULTIPOINT-TO-MULTIPOINT SERVICES

Karaoke on Demand (KoD) is the networked version of Karaoke. After a few drinks, you pick a song. The music, lyrics, and possibly visuals are streamed to your local karaoke unit. With KoD, it would be possible to select tunes from a large centralized library.

Moreover, it would be possible to have participants in other locations join in to create a virtual chorus! Lyrics could be available in a number of languages, and, in the case of a single participant, it would be possible to select tempo, reverberation, octave, background music, and background scenes. ISDN Karaoke on Demand in Japan is a good example of an actual deployment.

Although it is not the strongest imaginable market driver for residential broadband, KoD raises some interesting issues. First, it is a case in point that entertainment is something people will pay for—karaoke machines sell well, not only to drinking establishments but also to high-end residential users. Second, KoD represents very challenging technical problems, especially with regard to synchronization and conferencing, which require attention for business conferences and VoD.

Most importantly, the virtual chorus represents an application that requires a multipoint-to-multipoint topology. Other such applications are interactive gaming and video-conferencing, both of which have consumer applications. Gaming is an obvious example. A familiar interactive game is *Doom*, in which players at multiple sites stalk and shoot each other in a virtual combat zone. Video-conferencing also is a way for families to celebrate special occasions together, even when family members are geographically distant.

Multipoint-to-multipoint applications raise a particularly difficult quality of service problem. A single receiver might receive input from multiple sites simultaneously, producing congestion at the receiving site. Techniques are required to arbitrate the congestion and yet provide some bandwidth to each sender. This is a work in progress.

NARROWBAND SERVICES

A number of important and familiar narrowband services will coexist with RBB networks. Among these are utility metering and telephony.

Real-Time Utility Metering

Utility metering in real time over a network to the home has no direct value to the consumer. Its value is to the utility company, who will be able to:

1. Reduce the cost of invoicing customers.
2. Change pricing policy for utilities of all types, such as water, gas, and electric.

The first reason has relatively little appeal because it costs less than a one dollar a month to read a meter. The second reason, however, is more compelling.

In real-time metering, the utility company measures usage of electricity, gas, or water based on fine-time increments, perhaps as little as 15-minute intervals. This would enable the utility company to set prices based on peak usage, which in turn would motivate the consumer to use less of a utility during peak hours. Real-time metering of water, for example, would encourage consumers to wash clothes at night in the summer. The effect of real-time metering of electricity should be to flatten the demand for power during the day, while the total amount of power consumed stays the same. Flattened demand, as opposed to peak usage spikes, lessens the probability of power brownouts. Ultimately, this means that less capital is required to expand capacity, which is a big win for the utility company and the consumer.

Independent of real-time metering, the Telecommunications Act of 1996 permits utility companies to enter the data communications business. Because these companies have an incentive to create networks for real-time metering, it is logical that they can leverage these meter-reading networks into Internet-access networks.

One way for an electric utility to enter RBB businesses is to build new networks by using utility rights-of-way and installing new fiber. Utilities—especially electric utilities—have a particular advantage in this area because they certainly do not have to worry about remote powering of their routers and switches. In addition, electric companies have one of the prime ingredients of good network design—real estate. Utility poles and various rights of way permit the quick rollout of ubiquitous networking. Finally, meter reading is a very low bit-rate service. As little as 10 bps will do because a meter doesn't have much to say. That leaves plenty of bandwidth left over for other things.

Thus, utility companies, such as cable operators and broadcasters, have their own special incentive to offer residential networking. In the case of utilities, the incentive of real-time metering presents a unique financial incentive.

Telephony

Telephony, along with television, is the most familiar and important network service consumers have today. Telephony is a $650 billion annual service business worldwide (with billions more spent on equipment), thereby dwarfing both television broadcasting and data services. More will be said about the structure of telephone service in Chapter 4.

For now, the impact of telephone service on RBB networks takes one of two forms. The new high-speed networks can either ignore voice service entirely or embrace it.

One reason for not offering voice service is to avoid the regulatory issues associated with universal access service. Universal access refers to the public policy of making telephone service and certain utilities available to all consumers regardless of geographical location and even, to some extent, regardless of capability to pay. An even more encompassing reason for RBB networks to shy away from providing telephone service is the overall difficulty of the business—technical barriers and costs to new voice service providers are considerable. Finally, a general uncertainty surrounds how consumers will respond to having a multitude of voice service providers from which to choose. Although customers have shown a willingness to change their long-distance telephone operators, it is less clear that they will do the same for their local service providers. For these reasons, broadband networks such as DBS service so far are not involved with telephony at all.

On the other hand, new broadband networks might try to carve out a piece of the big telephone revenue pie, which is bigger than television or Internet access. Cable operators, for example, have a $25 billion business in the United States and look longingly at the $180 billion telephone business industry.

Some telephone services, particularly facsimile transmission services, are suitable for new networks. Fax messages, for instance, can be transmitted over the Internet. These fax transmissions account for large percentages of long-distance connections, especially for overseas voice calls. Fax messages, however, are not lifeline services and do not require real-time service. RBB networks easily could offer such peripheral services, revenues for which are now enjoyed by telephone companies.

For new networks that opt to offer voice service, the challenges will not include bandwidth limitations. Even if all the voice traffic in the United States was added together at any one instance, the volume of bits of data would still fall within reason. Nearly 100 million households exist in this country and about 20 percent of these have second lines. About 170 million telephone lines exist nationwide, including businesses.

If at any given moment five million voice calls are placed in the United States, connecting 10 million people at 64 Kb per connection, this results in 320 billion bps of traffic. Modern fiber transmission permits 40 billion bps on a single strand of fiber. Because fiber is being installed everywhere by telephone companies and by cable operators, surprisingly small amounts of their available bandwidth will accommodate the nation's telephone traffic.

The hard part of telephony therefore lies elsewhere. The difficulty lies in signaling (which is the intelligence part of the network), in service (which is the human part of the business), and in reliability, especially in times of need and on Mother's Day. Critics argue that Telecom 96 does not foster sufficient competition in the telephone service arena. With or without the Act, telephony is a hard business to enter and run. In summary, new RBB networks that intend to offer telephony will find it a bit more difficult than picking up a phone.

Narrowband's Relation to RBB

Narrowband services such as real-time metering and telephony are important because they represent a lot of money spent by consumers. Telephone companies and utilities each generate more than $100 billion a year in consumer revenue. A key question in the evolution of RBB is to identify the revenue split between narrowband and broadband services. Will they compete for the same pie, or will RBB need to uncover new revenue sources? Either narrowband services will be embedded in the new RBB networks and thereby help fund RBB rollout, or else RBB services will uncover new markets.

▶ AT ISSUE

Universal Access to RBB

Another important question posed by narrowband services is the role of universal service in the RBB environment. Telephone companies and utilities are accorded regulatory protection in exchange for guarantees of providing universal access. It is the public policy of this country to promote the widest possible distribution of these technologies, even if it requires cross-subsidies.

One openly accepted form of cross-subsidy is the perpetuation of the Universal Access Pool. The *Universal Access Pool* is a fund to which telephone operators contribute, for the purpose of defraying costs associated with rural telephony service. Rural telephone service providers tap into the pool to cover expenses under appropriate and stringent rules so that their customers can have telephone service and pricing comparable to urban service. A form of cross-subsidy that is not accepted by law is the use of funds from regulated telephone service to fund new, unregulated businesses, such as RBB. This form of cross-subsidy is viewed as unfair to competitors of telephone companies and is forbidden. Telephone companies historically have been obliged to create separate subsidiaries in which to incubate new businesses.

In May 1997 the FCC issued rules on the Universal Access Pool, pursuant to law in the Telecommunications Act of 1996. The rules provide for a contribution of $25 billion over five years to be funded by local exchange carriers, long-distance carriers, and cable operators who offer telephone service. These guidelines also specify how telephone service providers (mostly rural telephone companies) can take funding from the pool. Rural telephone companies, of course, wanted a larger pool, while urban telephone companies wanted a smaller pool. Internet service providers are exempt from contributing to the pool, but if telephone service were to be offered over the Internet, this clause would be revisited.

The issue of RBB information haves and have nots will continue to be subject to regulatory scrutiny.

RECENT U.S. GOVERNMENT ACTIONS

The following list details the key actions by the government in 1995–1996 that had implications for RBB:

- Spectrum auctions for wireless voice and paging services

- Government actions on digital TV

- Telecommunications Act of 1996

Together, these three actions emphasize the trend toward telecommunications deregulation and a commercial-based access to public infrastructure, such as airwaves. The investment community believes this trend is generally favorable to RBB deployment in the long run. The short-term effects, however, of Telecom 1996 have been negative (see the Viewpoint sidebar, "Wall Street's Reaction to Telecom 96," later in this chapter).

Spectrum auctions for digital telephone services convinced Congress that auctioning spectrum was a good thing—the FCC raised $9 billion, far more than anticipated. This was a relatively painless way to raise money. The impression created by the auctions is that the government is taking a more businesslike approach to spectrum management. Market forces will decide which services get how much bandwidth, rather than simple licensing and allocation of spectrum. The investment community is generally supportive of this approach, and capital has been raised for new services offered over auctioned spectrum. (If the notion of free digital spectrum for broadcasters seems to conflict with the value to the public of spectrum auctions, recall that it is Congress's intent to preserve free, over-the-air television.)

The CICATS/ACATS/Grand Alliance accord and the announcement by the FCC of a transition timetable to digital TV showed that the FCC encouraged the convergence of the video and data worlds. The United States is on a firm course with this convergence, and members of the video and data industries ignore each other at their financial peril.

The Telecommunications Act of 1996 was the first major piece of telecommunications legislation since the landmark Telecommunications Act of 1934. The goal of this new act was to recognize technical changes in various industries related to telecommunications and, accordingly, bring regulation up to date. In particular, the act acknowledges the blurring of technologies and the corresponding blurring of businesses that use those technologies. In most respects, the act is deregulatory, but in a few ways, strong regulation was enacted, particularly with respect to content.

The language of Telcom 96 is addressed directly to the broadcast, cable, and telephone service providers.

The Broadcasters

Telecom 96 provides for greater broadcast spectrum flexibility. This means that broadcasters can more or less use granted spectrum however they please.

As noted earlier in this chapter, ownership rules have been greatly relaxed for broadcasters as a result of Telecom 96. In addition, cable operators are now permitted to own local broadcast systems.

The Cable Operators

A majority of the Cable Act of 1992 was repealed by Telecom 96. In particular, rate regulation ends in March 1999. Rate regulation ends earlier for small operators or if a major cable operator can prove that a phone company provides video over a comparable number of channels.

Telecom 96 also allows cable operators to acquire up to 10 percent of small, local phone companies in their service areas.

The Local Exchange Carriers (LEC)

Local exchange carriers are providers of local telephone service. They are divided into two groups: *incumbent LECs (ILEC)* and *competitive LECs (CLECs)*. Familiar ILEC names are Pacific Bell, NYNEX, GTE, and independent phone companies such as Southern New England Telephone (SNET). The Telecommunications Act of 1996 itemizes ILECs. CLECs are entrepreneurs and big companies who currently compete or soon will compete for local phone service. These include wireless telephone operators and cable companies.

ILECs were granted immediate access to out-of-area markets for alarm, VoD, mobile communications, and data services. ILECs are barred from offering alarm services in area for five years, with the exception of a grandfather provision that allows Ameritech to offer these.

CLECs are permitted to enter the local telephone market and are assisted in doing so by language in the act that requires ILECs to offer interconnection services to the CLECs. State and local laws that impose barriers to unbundling, such as allowing for high resale rates of interconnection services, would be preempted by federal law.

VIEWPOINT

Trouble for ILECs

The ILECs are potentially the biggest losers of Telecom 96 because each ILEC has a duty to unbundle local service and make it available for resale. Interconnected carriers will connect to the ILEC network at technically feasible points—the meaning of this term no doubt will be settled in court.

continues

For a CLEC such as a cable company, for example, to offer telephone service means that it must connect to phone company networks to connect to phone company subscribers. The interconnection must be offered to the CLEC at discounts to be decided by regulations, some federal, and some state and local. The interconnection will yield a comparable level of service and technical features. Among these comparable features are access to space in the telephone company premises and dialing parity, meaning that the customer does not have to dial a cumbersome sequence of numbers.

Under current rules, ILECs may not include their marketing, billing, or collection costs in computing interconnection charges to the CLECs.

The entire apparatus of services and pricing are subjects of ongoing lobbying, which pits ILECS against CLECs.

The ILECs were not amused by the implications of Telecom 96. GTE, Bell Atlantic, BellSouth, Pacific Telesis, SBC Communications, and U.S. West filed a brief before the Eighth U.S. Appeals Court in St. Louis, asking the Court to vacate the FCC's entire unbundling provision. This stay was based on objections filed by two of the parties involved:

- ILECs claimed that unbundling of local loops is confiscatory, particularly because no agreement exists on accounting rules.

- State regulatory commissions objected to federal preemption rules, which they felt abridged their authority unconstitutionally.

The plaintiffs were granted their stay, but no doubt the matter will go to the Supreme Court for final resolution.

In other matters, the Telecommunications Act of 1996 permits the ILECs to enter the long-distance business (but only if an interconnect agreement with an alternative carrier for local loop competition exists) and to manufacture equipment (through separate subsidiaries). The ILECs also can provide video services and acquire stock in small cable companies in their service areas.

Utility Companies

As mentioned earlier in this chapter, Telecom 96 grants utility companies entry to data communications businesses through separate subsidiaries. This adds a very deep pocketed entrant to the RBB market. The utility companies are subject to oversight from the Securities and Exchange Commission.

Content

Telecom 96 made the transmission of indecent materials illegal, but the provision was overturned by the Supreme Court in June 1997. The response to this from the Internet industry was supportive, but the industry recognizes its public relations problem and intends to address it through technology and business practices. Certain companies, for example, such as SafeSurf (Los Angeles), produce lists of inappropriate web sites for children. ISPs purchase these lists and use them to bar access to those sites at a parent's discretion.

Additionally, Telecom 96 requires ratings for televised material similar to the ratings system of the motion picture industry. An advisory committee, consisting of members of the Motion Picture Association of America, National Cable Television Association, and National Association of Broadcasters, was established for the purpose of defining such a system; the proposed system is now in place.

The proposed system, however, is still under review by the FCC as this book goes to press. The system has been attacked by decency advocates who say it doesn't go far enough to protect children. It is unclear if providing more information about the content of a show will change viewing habits.

Related to the ratings system, Telecom 96 requires a technical means for parents to lock out programs with unacceptable ratings. This is the *V chip*, which links the ratings system to hardware; that is, the necessary circuitry must be built into the television. Thus, a final rating system must be in place before TVs with the appropriate hardware can be manufactured. Telecom 96 stipulates that the V chip is required by January 1998 for TVs larger than 13 inches.

VIEWPOINT

Wall Street's Reaction to Telecom 96

It's too early yet to know the full long-term impact of Telecom 96, but one effect is clear. Stock market valuations of LECs, cable operators, wireless operators, and long-distance carriers are down or have underperformed the market in the year since passage. Wasn't deregulation supposed to attract capital?

Telecom 96 encourages the FCC and state regulators to reduce prices and boost competition. If such a trend continues, where will the funds be found for network upgrades, particularly by telephone companies? Why should a telephone company invest in local loop technologies (one method of RBB delivery) if it is forced to resell services at prices it cannot dictate? If the ILECs can't invest in RBB, what is the incentive for cable companies to compete and upgrade significantly faster than the telephone companies? The reason for relatively weak telecommunications stock prices, in the greatest bull market of all, could be that Wall Street realizes that forced unbundling, deregulation, and competition make uncertain the prospects for a facilities-based infrastructure.

The ILECs are particularly defensive. Competitors and resellers will be permitted access to their core asset, the local loop. So far, some of their competitors, such as resellers, are not required to contribute to the universal access pool. The ILECs will have cost-based profit margins, while competitors and resellers have no profit restrictions. No wonder the ILECs are in court.

Clearly, the old Telecommunications Act of 1934 no longer makes sense in a world where phone service can be delivered on a host of networks. A rewrite was necessary, and considering the contentiousness and dollars involved, a reasonable compromise was reached. Undoubtedly, there will be growing pains as all parties settle into the new regulatory regime.

FUNDING MODELS

Where will the money come from for RBB networking deployment? Billions of dollars will be required in central sites for the production and distribution of content, for networking infrastructure, and for consumer electronics devices.

The following four classes of funding can be considered.

- Work-at-home subsidies

- Subscription

- Transaction

- Advertising

Each of these classes will be discussed in the following sections.

Work-at-Home Subsidies

Businesses (or the government, through tax deductions) often pay for some high-tech consumer products for business use. This was true of personal computers, fax machines, pagers, and ISDN, all which originally were paid for by companies for use by their employees at home. In turn, the companies wrote these purchases off as business expenses, so the taxpayers ultimately picked up part of the tab. By using this method, high-tech devices in the home in effect are not paid for by the early adopters of the technology.

This funding model benefits product development by transferring the cost away from the consumer. Stimulated by work-at-home subsidies, a market might become large enough to drive manufacturing and sales volume to the point where prices are reduced for purely consumer applications. The questions for RBB are whether it will be subsidized in this way initially, and whether its work-at-home applications eventually will be extensive enough to trigger reduced costs to consumers.

Although there was some hobbyist interest in the personal computer in the early 1980s, for example, the personal computer market was largely a business phenomenon during its early years, spurred on primarily by the development of spreadsheets and word processors. (Remember Wordstar and VisiCalc?) The scale of the business market for PCs became large enough to reduce costs for major PC components so that PCs came into reach of work-at-home subsidies. Separately, Apple Computers, realizing that home computing was financially out of reach for a mass market, chose to address the educational market. By servicing markets as large as business and education, economies of scale could be achieved so that component costs were reduced. Vendors reduced prices even more to expand markets even further by trimming features from commercial offerings and adjusting their distribution and support services.

The analogous situation for RBB is that the work-at-home employee will use RBB services for business purposes and have the company defray the expense. After hours, the employee will use the same services for personal use.

Subscription Revenues

Subscription revenue refers to the periodic income received from consumers in exchange for an ongoing service. Subscriptions for most consumer services—such as newspapers, magazines, cable TV service, and basic telephone—are in the relatively tight range of $10–$30 per month. Consumers have a limited tolerance for regular charges, and marketers target the pricing for this range. Basic RBB services will fall into a similar range.

Transaction Fees

In most media today, subscription revenues alone do not generate a profit. Whether it is newspapers, magazines, Internet access, or monthly cable charges, subscription revenues must be augmented with transaction or advertising revenue to make the business profitable. The same is likely to be true for RBB.

Transaction fees are charges paid per specific event or transaction, such as payments for VoD or PPV screening, fees paid to receive information off the web, or surcharges paid for home shopping. Transaction fees represent a potentially enormous source of revenue to content and service providers. Roughly $500 million of online spending occurred in 1996 (with 10 percent of web merchandising being sex-related).

Recent successes of a less-prurient nature include Amazon.com, a company that sells books over the Internet and recently enjoyed a successful initial public offering (IPO).

Companies such as CyberCash are built on the business premise that large-scale shopping will take place over the networks and that transaction fees will be substantial. The scale of transaction fees is unknown for now, but it is clear that such systems require detailed administrative systems that largely are not in place.

Advertising

Television advertising is a $34 billion business in the United States alone. Total U.S. advertising (newspapers, magazines, and broadcast) was $173 billion in 1996 and is increasing at five percent per year, which exceeds the growth rate of population and the Gross National Product (GNP). Even newspapers, thought by many to be outmoded in this electronic age, collect upward of $37 billion in the nation for print advertising and classified ads.

Can RBB and Internet access command a large enough following to generate advertising revenue on the level of television advertising? How large must that audience be? The answers to these questions for the moment are "Probably" and "Not sure, but a lot more than the 18 million online today." Heavy hitters, however, are placing their bets now that digital storytelling will generate lots of new advertising opportunities. The Creative Artist Agency (CAA), Cicso Systems, and Intel have built a digital studio in Los Angeles. Microsoft is spending $400 million a year on MSNBC and other media ventures, and Time Warner is spending $10 million on its web site, Pathfinder, looking for clues on how to create an advertiser-friendly environment.

When dividing TV advertising by households, U.S. advertisers spend about $340 per household per year to get their collective messages out. (The rest of the world spends about $157 per year.) This exceeds what most projections are for annual Internet subscription fees or annual basic cable TV fees. Advertising pays, especially for consumer markets, and it will continue to be profitable.

Web advertising can be sold in ways similar to print and television advertising. A web site, for example, can charge rates based on the size of an ad (such as a full page or quarter page) for a specific period of time (such as a week or a month), with rates varying by the expected number of visitors to that site. The ad buyer has no guarantee, of course, that anyone will read the ad, much less buy the product because they read the ad. So the ad buyer accepts the risk and the web site accepts the money.

The Internet permits a variety of advertising models beyond those available to print publishing or broadcast TV software tools, for example, exist to count *impressions*, that is, the number of times a web page is seen. Another parameter that can be measured is *page views*, which is the number of times a viewer clicks an icon and downloads a full page of graphics and text. Both impressions and page views represent methods of charging for web advertising that are unique to the medium.

Impressions are incremented when a viewer accesses a web page and when a banner or other graphical element containing advertising is displayed. Popular web sites can charge for impressions aggressively because of their high volume of traffic. Advertisers are charged for a specific number of impressions, say 1 million, at $30 per thousand, a rate termed CPM (cost per mill). CPM is established advertising jargon, familiar to ad buyers and agencies.

Page views are incremented when the viewer clicks the banner impression for further information. Page views represent a higher level of interest by the viewer in the advertising; as such, counting page views is a significant tool for evaluating ad effectiveness. Only a complete download is counted as a page view.

If an impatient viewer aborts the download prematurely, perhaps by returning to the previous screen, that download doesn't count as a page view. Web sites that charge on the basis of page views would like to prevent aborted downloads. Such sites would benefit from RBB because this technology provides much faster web retrievals, thus increasing page views and pleasing advertisers.

Some advertisers will pay only for a fixed number of downloads, regardless of how long it takes to reach that number. It would, for example, be possible under some pricing schemes to pay for 50,000 page views, irrespective of the length of time the ad is on the web site. If no one comes to the site, or if downloads are aborted prematurely, the amount of time the ad is available on the web page is lengthened. In this scenario, the advertising risk shifts to the web site and away from the buyer.

A problem with counting the number of impressions or page views is the use of software robots that roam the Internet to index web sites. These are pieces of software that set out to find sites on the Internet, looking for keywords so that search sites can inventory them. When a robot hits a web page, it increments the number of impressions or page views, which, of course, does the advertiser no good at all.

Because consumers' buying habits in response to web advertising are unknown, ad buyers sometimes prefer to pay for ads based on a percentage of revenue they derive from web sales. This can be easily tracked because the order process is handled through the web site. This appears to be a growing trend as the balance of power in Internet advertising moves from the web page to the advertiser.

Internet ads now range from $2,000–$30,000 per month for banners. The more expensive pages to advertise on are search engines (among the most popular web sites on the Internet), high technology pages, and adult-oriented pages.

The Internet Advertising Bureau reported $267 million in Internet ad spending for the 1996 calendar year. A report issued by media consultants Cowles/SIMBA Information estimates that the calendar year 1997 figure will be $446.2 million, growing to $2.5 billion by 2000. More than half of current spending comes from only 17 web sites, mostly search engines such as Yahoo!.

Advertising has the potential to produce the greatest amount of funding in the shortest period of time. Entertainment is a key RBB market driver, not only because it commands a significant audience but because it also provides the venue for delivering advertising. Even with ratings of the national broadcasters sagging (down to 64 percent for the four major networks in spring 1997, as opposed to more than 75 percent in 1990), their ad rates continue to hold firm because they remain the only means to reach a mass audience quickly in an increasingly fragmented advertising market.

SUMMARY

Consumers worldwide have high utilization of telephone and television services over existing telephone, over-the-air, and cable networks. For the most part, the services are of acceptable quality with a reasonable price.

To justify the multibillion-dollar rollout of RBB services to the home, new services will be required to stimulate consumers or advertisers to underwrite the costs. Table 1–7 summarizes the market drivers that are fulfilling or could fulfill this role.

RBB is interesting work, which is not the least of the reasons why it is evolving. RBB is attracting top-flight talent in entertainment and engineering. The next chapter examines the engineering challenges and technical foundations, in various stages of solution, of RBB. The remainder of this book frames the discussion on service evolution and delivery. These chapters explore what pieces of the RBB network puzzle exist and what pieces are missing so that readers can evaluate who the winners and losers will be.

Table 1–7   *Summary of Market Drivers for RBB*

Service	Characteristics	Key Challenges and Disincentives	Key Benefits and Incentives
Digital TV	Point-to-multipoint Push mode	Cost of infrastructure overhaul	Offers more channels Offers better picture Offers more programming options Builds on ubiquitous consumer demand for existing service, analog TV
Virtual channels	Point-to-multipont Push mode	Involves identifying appropriate content Requires processing capability at the consumer site; could be expensive	Reduces bandwidth requirement to transmit certain types of programs
Video on Demand	Point-to-point Pull mode	Cost of royalties, networking, and storage High bandwidth requirements Content availability	Offers a potential scale of market that is the same as home video (sales and rentals) Affords high consumer control Builds on consumer interest in videocassette rentals
Near Video on Demand	Point-to-mulitpoint Push mode	Consumer interest questionable Royalties Channel availability from cable and DBS	Builds on consumer interest in PPV
Web	Point-to-point Pull mode	Difficulties in scaling point-to-point networks Uses point-to-point access, so had difficulty attracting advertising revenue	Builds on consumer interest in the web Facilitates transactional revenues and online shopping

Table 1–7 *Summary of Market Drivers for RBB, Continued*

Service	Characteristics	Key Challenges and Disincentives	Key Benefits and Incentives
Push mode data	Point-to-multipoint Push mode	Consumer interest questionable Channel capability of cable and satellite restricts offerings, for now	Builds on consumer interest in stock quotes and information services Builds on carrier interest in software and data distribution
Video-conferencing	Multipoint-multipoint Push mode	Very complicated systems design; requires multipoint bandwidth guarantees Consumer equipment nonexistent	Offers completely new forms of personal entertainment Builds on consumer interest in visiting friends and relatives; access provided for shut-ins
Telephony	Point-to-point	Scaling	The mother of markets
Utility metering	Point-to-point	No utility company marketing of data and entertainment services; must be justified on basis of cost reductions	Real-time metering motives flatter consumer usage, which saves capital expenses Carries minimal bandwidth requirements Leverages widespread presence of utilities

REFERENCES

Books

[Brinkley], Joel. 1997. *Defining Vision: the Battle for the Future of Television.* Harcourt Brace, ISBN 0-15-100087-5.

Claiborne, David. 1990. *Mathematical Preliminaries for Computer Networking.* John Wiley Press, ISBN 0-471-51062-9.

Keith, Jack. 1993. *Video Demystified: A Handbook for the Digital Engineer.* High Text Publications, ISBN 1-878707-09-4.

Poynton, Charles. 1996. "A Technical Introduction to Digital Video." John Wiley, Second Edition, ISBN 0-471-12253.

Vogel, Harold. 1995. *Entertainment Industry Economics.* Cambridge University Press, Third Edition, ISBN 0-521-47070-6.

Watkinson, John. 1996. *Television Fundamentals.* Elsevier Press, ISBN 0-240-51411-4.

Articles

"Attendee Handbook." 1996 Digital Video Test Symposium, Hewlett-Packard.

"Communications." *IEEE Spectrum*, January 1997, pages 27–37.

"Consumer Electronics." *IEEE Spectrum*, January 1997, pages 43–48.

"Digital TV: Cost Risk for Pure Play Broadcasters." *Wall Street Journal*, January 2, 1997.

ETSI ETS 300 468 "Digital Broadcasting Systems for Television, Sound and Data Services; Specification for Service Information (SI) in Digital Video Broadcasting (DVB) Systems."

[Grand Alliance]. "The U.S. HDTV Standard." *IEEE Spectrum*, April 1995, pages 36–45.

Grossman, Dan. "General System Requirements for RBB." *Document Number: ATM Forum/95–1047*, August 1995.

Petajan, Eric. "The HDTV Grand Alliance System," *IEEE Communications*, June 1996.

Pugh, William and Gerald Boyer. "Broadband Access: Comparing Alternatives." *IEEE Communications*, August 1995.

Stordahl and Murphy. "Forecasting Long-Term Demand for Services in the Residential Market." *IEEE Communications*, February 1995, pages 44–49.

[Wiley], Richard. "Digital Television: The End of the Beginning." *IEEE Communications*, December 1995, pages 56–57.

Internet Resources

http://www.atsc.org/
(The web site of Advanced Television Systems Committee [ATSC], which summarizes the digital TV agreement.)

http://www.cs.cmu.edu/afs/cs.cmu.edu/user/bam/www/
numbers.html#OtherSources
(A quick source for computer numbers.)

http://www.fcc.gov/Bureaus/Mass_Media/Notices/fcc96317.txt
(FCC DTV channel allotments as of the summer of 1996.)

http://www.fcc.gov/Bureaus/Mass_Media/Orders/fcc96493.txt
(FCC statement on the ACATS/CICATS compromise, December 26, 1996.)

http://www.fcc.gov/oet/headline/fcc97115/Welcome.html
(FCC DTV channel allotments as of April 1997.)

http://www.fcc.gov/Speeches/Ness/spsn621.txt
(The FCC announcement of the Grand Alliance/CICATS accommodation.)

http://www.gi.com/products/DCSYSTEM.htm
(GI on MPEG and DigiCipher, their proprietary compression scheme.)

http://www.ipmulticast.com
(The home of the IP multicast Initiative.)

http://law.house.gov/cfrhelp.htm
(*The U.S. House of Representatives Internet law library.*)

http://www.mindspring.com/~kjack1/video_standards.html
(*Video standards around the world.*)

http://www.nortel.com
(*Use Nortel's search engine to find Telephony 101. Northern Telecom has good White Papers.*)

http://www.openmarket.com/intindex/
(*Fun Internet statistics thanks to* Harper's Magazine.*)*

http://www.research.microsoft.com/research/graphics/Alvy/DigitalTV/Digital%20TV.htm
(*A view of the digital TV process from Microsoft.*)

http://www.tek.com/VND/Support/#white
(*Video and Networking White Papers by Tektronix.*)

http://www.whitehouse.gov/fsbr/ssbr.html
(*Up-to-the-minute social statistics from the White House and the Census Bureau.*)

CHAPTER 2

Technical Foundations of Residential Broadband

Residential broadband (RBB) service requires an understanding of a variety of technologies. Many important technologies, such as *ATM (Asynchronous Transfer Mode)*, have been treated extensively elsewhere, and are relatively free of controversy, and well understood. Other topics, however, are less familiar. This chapter presents a selection of technical foundations based on at least one of these qualifiers:

- The topics mention in the trade press as noteworthy or contentious issues

- The topics applicability to a wide range of RBB technologies

- The topics relatively light treatment elsewhere

This chapter begins by developing a reference model for an RBB network and then provides an overview of the characteristics, problems, and benefits associated with a number of important related technologies.

Many of the technologies discussed here will be revisited in more detail, as appropriate, in later chapters. This chapter's coverage is meant to provide a wide but fairly basic perspective on the technical foundations of RBB. If you are already familiar with these subjects, you might want to proceed directly to Chapter 3, "Cable TV Networks."

A REFERENCE MODEL FOR RBB

This section develops a generic RBB network by establishing a reference model that incorporates the key features of RBB networks. A *reference model* is useful in establishing vocabulary and pinpointing parallel features among various technical approaches. Finally, a reference model modularizes network functionality and helps identify points of congestion in end-to-end service delivery.

Useful reference models for RBB networks have been developed by the Digital Audio Visual Council known as DAVIC (http://www.davic.org) and ATM Forum Residential Broadband Working Group known as ATMF RBB (http://www.atmforum.com). The DAVIC Reference model is written in DAVIC 1.0 Specification Part 02, "System Reference Models and Scenarios," September 1995. The ATM Forum Reference model is presented in the document BTD-RBB-01.01, "Residential Broadband Baseline Document Draft," February 1997.

This discussion focuses on the DAVIC model because this particular model provides greater generality. Although many components of RBB will use ATM, it is not necessarily assumed that RBB is ATM-centric.

Figure 2–1 depicts the end-to-end schematic for the reference model. As illustrated, an RBB system consists of content providers, a Transport Network, an Access Network, and a Home Network system. The domains are segmented in this way to emphasize boundaries of control. Splitting the content provider network from the Transport Network, for example, illustrates that these two networks are run autonomously and therefore require some interface (which DAVIC refers to as the A9 interface) to provide a standard between the two networks.

Note that although this discussion borrows heavily from DAVIC, it varies somewhat from the formal DAVIC model referenced previously. First, DAVIC does not explicitly mention the *optical network unit (ONU)*. The ONU is included here to emphasize the important role that fiber optics plays in RBB. In some cases, the ONU is located inside a large telephone company central office. In other cases, it is located near the customer's home, for example, on a street corner or a telephone pole. The placement of the ONU is a recurring theme for all Access Networks. In addition, DAVIC makes no mention of a Residential Gateway, which is explicitly defined here and described further in Chapter 7, "In-Home Networks." Finally, the DAVIC interfaces A3 and A4–A8 are internal to the Access Networks and Transport Networks and are useful primarily to carriers implementing them. Therefore, this chapter will not refer to these interfaces.

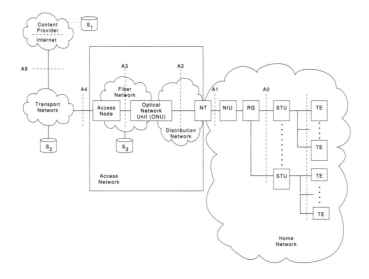

Figure 2–1

Reference model for RBB network.

Courtesy of the Digital Audio Visual Council (DAVIC) (http://www.davic.org)

Content Provider Network

Content emanates from a server, S1, on a *content provider network*. Content providers own content or have licenses to distribute content such as movies, web server information, and copyrighted printed matter. The content provider network can be the Internet, a corporate network, or some application-specific network. It may or may not utilize IP, but it probably connects to a core Transport Network operated by a large carrier with ATM. Therefore, interface A9 is likely to be ATM. ATM allows content providers to use a single, high-capacity connection to the Transport Network, while controlling quality of service to individual subscribers.

Transport Network

The *Transport Network* provides high-performance bit pumps to move bits between the content provider and the Access Network. This network is the repository for network-management and application servers, depicted in Figure 2–1 as S2. Speed and manageability are critical in the Transport Network. The following list details the principle functions of the Transport Network:

- Switching, routing, and transmission
- Concentration of traffic to the content provider
- Service registration for multiple content providers
- Multiplexing and switching for multiple Access Networks
- Navigation aids and directory services
- Quality-of-Service enforcement
- Load balancing

- Caching servers on behalf of the content providers

- Replication for multicast and broadcast services

Some of these functions can be performed by the Access Network as well.

Conventional wisdom is that the core of the Transport Network will be provisioned with high-speed ATM fiber-optic links. Video or web servers will be connected with ATM, Fast Ethernet, or Gigabit Ethernet interfaces to switches and routers in the Transport Network. Other options to consider in the Transport Network are packet mode over Sonet and circuit switching for some point-to-point transmissions.

In Figure 2–1, the Transport Network connects to the Access Network over the DAVIC interface A4.

Access Network

The *Access Network* is the part of the carrier network that touches the customer's premises. The Access Network also is referred to as the local drop, local loop, or last mile. It connects to the network termination on the customer's premises over the DAVIC A2.

The following list details the principal functions of an Access Network:

- Transmitting, switching, routing, and multiplexing traffic from the consumer to the Transport Network.

- Classifying traffic from consumers by quality of service. The Access Network differentiates best-effort traffic from traffic with guaranteed bandwidth.

- Enforcing quality of service.

- Providing navigation aids and directory services.

- Caching servers on behalf of the content providers.

- Handling tunneling or packet encapsulation.

- Enforcing MAC protocol.

- Enforcing packet filtering.

- Updating software on the Home Network.

- Authenticating users.

- Establishing measurements used for invoicing.

These functions are performed cooperatively with the Transport Network, the content provider, and the Home Network, which means that clear interfaces must be established among these systems.

AT ISSUE

Network Intelligence

An important system design question is how network intelligence is to be optimally distributed between the Transport Network and the Access Network. Intelligence here refers to processing of user signaling (such as call-setup requests and changes of call state). [Quayle] argues that the Access Network should be relatively dumb and that the Transport Network should be relatively intelligent. Furthermore, large databases for invoicing and resource management should be maintained in the Transport Network. Because relatively few Transport-Network nodes exist and a large number of Access-Network nodes exist, total system costs are optimized by concentrating the expensive (intelligent) pieces in the Transport Network.

Others argue that because the Transport Network must move bits as quickly as possible, this network must be simple. Therefore intelligence is pushed to the periphery, even to the customer premises. ATM switches, for example, are relatively simple, with most of the difficult software functions performed outside the switching fabric. In addition, because some Access Networks require a media access-control protocol, there is a requirement to interpret user signaling in the Access Network. This is not the case for Quayle's employer, British Telecom.

Finally, on a commercial matter, it might be the case that the Access Network and Transport Network will be different companies. Each will assert that intelligence should be put into its network, which boosts added value.

Many types of Access Networks are discussed widely in the popular and trade press. Among these are those detailed in the following list:

- Hybrid fiber coax (HFC)

- Advanced digital subscriber loop networks, such as Asymmetric Digital Subscriber Line (ADSL) and Very high bit rate Digital Subscriber Line (VDSL)

- Fiber to the Curb (FTTC)

- Fiber to the Cabinet (FTTCab)

- Fiber to the Home (FTTH)

- High-speed wireless, such as Multichannel Multipoint Distribution Service (MMDS) and Local Multipoint Distribution Service (LMDS)

Each of these networks will be discussed in subsequent chapters.

Although multiple types of Access Networks exist, they all share three architectural elements: the access node, the optical network unit (ONU), and the network termination (NT in Figure 2–1).

Access Node

The *access node (AN)* performs some or all the following functions:

- Modulating forward data onto the Access Network

- Demodulating return-path data

- Enforcing the *Media Access Control (MAC)* protocol for access onto the Access Network

- Multiplexing traffic from the Access Network to the Transport Network

- Separating or classifying traffic prior to multiplexing onto the Transport Network, such as differentiating traffic that is subject to Quality of Service (QoS) guarantees from traffic that receives best-effort support

- Enforcing signaling

- Handling passive operations, such as splitting and filtering

Enforcement of the MAC protocol is one of the more complicated functions of the AN. When multiple consumers vie for bandwidth connections simultaneously, some Access Networks use arbitration rules to determine who gets to use the network at that moment. Without some controls, the Access and Transport Networks would be obliged to provide an indefinite amount of bandwidth and connections. The MAC protocol arbitrates access to the Access Network. This is enforced by the AN, which engages in a protocol exchange, or peers, with the NIU on the customer premises. Important MAC protocols are token-passing schemes, Slotted Aloha protocols, and Collision Sense Multiple Access protocols.

Examples of ANs are Cable Model Terminal Service (CMTS) in the Multimedia Cable Network System (MCNS) architecture for cable data modems; Digital Subscriber Line Access Multiplexer

(DSLAM) for ADSL concentration; and the ATM Digital Terminal (ADT) referenced in the ATM Forum RBB baseline document.

Optical Network Unit (ONU)

At this point, the reference model departs somewhat from the DAVIC model by specifying a network component called the *optical network unit (ONU)*. The ONU is understood to be part of all proposed Access Networks but does not have an explicit definition in DAVIC.

The function of the ONU is to terminate fiber and convert the optical signal on fiber to electrical signals on wired networks. A major difference among Access Networks is the proximity of the ONU to the consumer. The closer the ONU is to the consumer, the higher the speed and usually the greater the service reliability. Carriers and consumers alike prefer to have fiber as close as possible to the home—perhaps even inside the home, in the case of service called Fiber to the Home (FTTH).

Network Termination

The *network termination (NT)* is a carrier-provided or customer-provided piece of equipment that is located on the side of the home. The following list details some of the NT's functions:

- Coupling of home wiring to the carrier wiring
- Grounding
- RF filtering
- Splitting
- Media conversion
- Remodulation

- Security and interdiction

- Provisioning

- Loopback testing by the carrier

The NT is the legal and commercial demarcation point between the Access Network and the consumer. This means that if something goes wrong on the network side of the NT, the carrier is obliged to fix it. The NT is situated on the customer premises, but DAVIC includes it as a part of the Access Network.

In the United States, the NT is considered to be customer-premises equipment because the customer purchases it. In Europe, the NT is considered carrier property. For purposes of this reference model, the NT is part of the carrier network because it usually is under carrier control, even though the customer purchases it. The fact that NT can be purchased from a number of suppliers other than the access carrier is not relevant architecturally.

NT functions differ widely among Access Networks, but the main functions are usually passive (such as coupling, splitting, grounding, and RF filtering). In some cases, however, the NT can have an active function, such as signaling. Service activation, for example, might be enabled by signaling between the carrier and the NT. This will become an important concept for FTTH, which is discussed in Chapter 4, "xDSL Access Networks."

Chapter 7, "In-Home Networks," contains more discussion of various Home Network components.

Why Multiple Access Networks Exist

Proponents of specific Access Networks claim that one Access Network would fit all applications. Although it is technically feasible that a single network could service all market drivers, the following are reasons why multiple Access Networks do exist:

- **Embedded base.** Billions of dollars have been put into the ground by telephone companies and cable operators. This represents investment that cannot be easily replaced. The sensible thing is to leverage it.

- **Different applications.** Broadcast television lends itself economically to shared media, such as air and cable. Retrofitting broadcast media to accommodate point-to-point applications involves software and hardware tricks. Similarly, retrofitting broadcast applications over point-to-point media also involves software tricks.

- **Different population densities.** Some technologies are more economic in urban areas than rural areas, and vice versa.

- **Different geography.** Vast expanses of oceans and rivers create a cost penalty for wired networks. On the other hand, obstacles such as hills, buildings, vegetation, and heavy rainfall reduce transmission quality for wireless networks.

- **Different business conditions.** In developing countries, which are characterized by a lack of mature wired infrastructure, broadcast technologies can create a fast infrastructure. Regulations such as those imposing limits on market dominance of a single carrier can foster or inhibit some technical options.

Because of these variations, it is worthwhile to consider a variety of Access Networks to understand the strengths and weaknesses of each. Chapters 3–6 will address the different networks.

Servers

A final point to make about Access Networks is that servers may be moved from the content provider premises or the Transport Network premises to the Access Network. In Figure 2–1, server S1 is replicated on the Access Network. This is to take advantage of the speed of the local loop. If the content or Transport Network is slow, the advantages of the fast local loop are lost to the consumer. Therefore, some access-network providers are providing web caching services.

Service Consumer System (SCS)

The *Service Consumer System (SCS)* is the DAVIC term for the in-home infrastructure for broadband networking. The SCS, according to DAVIC, consists of the following components:

- The *network interface unit (NIU)*, essentially a modem

- The *residential gateway (RG)*, which adds network functionality and multiplexes different services

- The *set-top unit (STU)*, which performs application-specific functions such as decoding digital TV

- The *terminal equipment (TE)*, which is a television, a personal computer, or another device

- Consumer premises distribution, either wired or wireless

The Residential Gateway (RG) is a new component not formally defined by DAVIC. The RG and other components of the SCS are covered in more detail in Chapter 7.

METALLIC TRANSMISSION MEDIA

The reference model provides a framework for considering some of the specific technologies on which RBB depends, as well as the challenges and benefits associated with those technologies. This section covers a very familiar, ubiquitous technology: metallic wiring. Fiber transmission is discussed later in this chapter, and wireless transmission characteristics will be discussed in Chapter 6, "Wireless Access Networks."

Problems Associated with High Frequency on Wires

Table 2–1 summarizes several important data services in terms of bandwidth and the type of metal wire used. The types of wire listed provide well-known, stable services for the indicated frequencies. But RBB intends to push frequency used on phone wire up to 10 MHz and frequency used on cable up to 1 GHz. This will be difficult, owing to impairments associated with high-frequency transmission on metal wire. The following list details some of these impairments:

- More attenuation

- More crosstalk

- More resistance (skin effect)

- More phase error

Table 2–1 *Representative Services and Bandwidth for Metal Wire Transmission*

Service	Upper-Limit Bandwidth	Type of Wire
Voice service	3,400 Hz	Phone
Alarm service	8,000 Hz	Phone
ISDN	80,000 Hz	Phone

Table 2–1   *Representative Services and Bandwidth for Metal Wire Transmission, Continued*

Service	Upper-Limit Bandwidth	Type of Wire
T1 using 2B1Q	400,000 Hz	Phone
Cable TV	350 million–750 million Hz	Coaxial cable

Attenuation

Signal loss, or *attenuation*, is a function of frequency. As frequency increases, the distance the signal can travel decreases by the square root of the frequency. This means that a 40 MHz signal will travel half as far as a 10 MHz signal. Velocity of the signal through a wire is another function of frequency. The higher the frequency, the slower the signal.

Because optical signals through glass fiber retain signal strength better than electronic signals through metallic wires, they provide improvements required for RBB services.

Crosstalk

When two adjacent wires carry signals, the possibility exists that signals from one wire will enter the other wire as a result of electromagnetic radiation. Crosstalk increases with increasing frequency. At low frequencies used by voice, crosstalk is not noticeable. But at frequencies required by high-speed services, it is a principle cause of signal degradation.

Resistance

As signals are transmitted through wires at very high frequencies, a phenomenon called the *skin effect* occurs. This is the behavior whereby electricity migrates to the outside wall of the wire, leaving little conductivity in the middle of the wire. In fact, in extreme cases it is possible to replace the core of the wire with cheaper nonconductive material, such as plastic or wood. As electricity migrates to the skin, resistance increases because less of the wire is used. This increased resistance weakens signals.

The skin effect explains why there are no services above 1 GHz over wired media, whereas the over-the-air spectrum can be used for frequencies in the range of 20–30 GHz.

Phase Error

Higher frequency signals not only weaken more quickly when compared with lower frequencies, but they also are a little slower, which causes a phase error. For modulation techniques dependent on phase, this can introduce bit errors.

External Impairments

Apart from inherent problems associated with high frequencies, metallic networks encounter external impairments. Many of these impairments have to do with noise, which is caused by disturbances that reduce the clarity of the signal being sent. A challenge for RBB engineers is to characterize noise for specific networks so that proper encoding and noise mitigation techniques can be used. Noise characterization involves determining whether impulse noise or narrowband noise is the major cause of imperfections, at what frequency the noise occurs, and what causes the noise. Noise characterization of cable TV networks continues to be the subject of research by CableLabs, a research consortium for North American cable operators located in Louisville, Colorado. Bellcore does the same for telephone services.

Leakage

Outdoor insulation suffers wear and tear. Deterioration of the foil jacket or cladding enables radiation from the conducting material to leak through the perforation to the outside. This is a source of signal loss. Moreover, the radiated signal that emanates from the wire can cause interference with wireless services, such as radio transmissions. This is a reason why the FCC has rules limiting signal leakage. Morever, leakage can come from inside customer premises equipment, such as television sets.

Another cause of leakage is poorly fitted couplings at the customer site, such as F connectors for cable TV not tightly joined to the cable box. This presents a system-management problem because such leakage falls outside the control of the carrier.

Impulse Noise

Impulse noise (or burst noise) occurs when one wire picks up an unintended signal that lasts for a short period of time (such as a few microseconds) but that interferes with a wide frequency range. Sources of impulse noise include other wires, motors, and electronic devices. Impulse noise generally is caused by imperfections in the wire, such as corrosion or malfunctioning amplifiers.

Narrowband Interference

Another type of noise is *narrowband interference*, which affects a small number of frequencies over a longer period of time. Amateur radio, or AM or FM radio interference, are types of narrowband interference. Consider AM transmission occurring at 1070 KHz. If a signal is at the same frequency in a wire, and a lesion is in the wire or a loose fitting, then receivers at the end of the wire will pick up the signal at 1070 KHz.

One way to avoid the interference is to identify the frequency range of the noise in advance and simply not transmit on that frequency. This is a technique that is used by the current modulation standard of ADSL, which is discussed in Chapter 4, "xDSL Access Networks."

Loading Coils

Loading coils are devices that lengthen the distance over which voice can travel across phone wire. They limit power at higher frequencies and facilitate energy to lower frequencies (less than 4 KHz) where the voice passband sits. The positive effect is that this provides better voice fidelity over a longer distance. The negative effect is that loading coils eliminates the use of high frequencies that are needed for broadband services. Loading coils are one factor that limits phone wire to narrowband services. This is hardly surprising, because loading coils were devised decades ago, before thoughts of high-speed services.

Loading coils preclude the wire from transmitting any digital service, such as ISDN, ADSL, or leased-line T1 service. The problem of loading coils is resolving itself slowly as the coils periodically are removed by the introduction of new services. Approximately 20 percent of local loops in the United States currently have these coils.

Bridged Taps

Bridged taps are wires that are connected to a network, in which one end of the wire is unconnected to proper termination—it's just dangling. This can happen when consumers or field technicians attach or remove devices without completely disconnecting wires to the old device.

The unterminated wires cause two problems. First, they cause leakage, which causes a loss of signal strength. Second, signals echo back off the unterminated end into the mainstream network, interfering with good signals.

Locating bridged taps to correct them can be difficult. Network providers often have incomplete records of where the taps occur, especially because bridged taps can be instigated by the consumer without the provider's knowledge. Thus, in both telephone companies and cable networks, bridged taps must be addressed through some kind of noise mitigation technique.

AT ISSUE

Legacy Wiring Challenges for Telephone Companies

A continuing concern among telephone companies worldwide is how much of their existing local loop plant is actually usable for high-speed networking. Age, rust, corrosion, broken insulation, loading coils, bridged taps, and occasional poor record-keeping about all these problems jeopardize the potential to provide RBB services over legacy wiring. The amount of wiring that can be redistributed for RBB makes a big difference in the financial spreadsheets for new services at the telephone companies. The more wiring that must be installed, the longer it will take to deploy, the more it will cost, and the greater advantage goes to telephone company competitors, such as cable operators. The condition of wiring also could have an impact on regulatory deliberations, as the telephone companies might seek some rate relief to upgrade legacy systems. The process of loop qualification must be watched closely to calibrate the near-term rollout of RBB services.

MODULATION SCHEMES

Modulation schemes are another central technology for RBB. Digital transmission over-the-air or through metallic wires (such as telephone wires or coaxial cable) requires the use of modulation schemes (also called line-coding techniques) to infuse digital information onto the medium. Modulation schemes differ as to the speed of service they provide, the quality of wire they require, noise immunity, and complexity. Modulation schemes are a hotly debated issue, and many incompatible schemes have developed. These schemes come in two broad classes: single-carrier techniques and multiple-carrier techniques.

Single-Carrier Techniques

Single carriers are so called because a single channel occupies all their bandwidth. Single-carrier modulation techniques permute any of three characteristics of an analog wave form—amplitude, frequency, and phase—to encode uniqueness, which is, after all, information. Frequency modulation is familiar for use in radio transmission but is not used in digital techniques, which use only combinations of amplitude and/or phase modulation.

2B1Q

An amplitude modulation technique used for ISDN and High bit rate Digital Subscriber Loop (HDSL) service in the United States is *2B1Q (2 Binary 1 Quaternary)*. This is defined in the 1988 ANSI specification T1.601.

2B1Q has four levels of amplitude (voltage) to encode 2 bits. Each voltage level is called a *quaternary*. Because of the four voltage levels, each level translates to 2 b/Hz.

Bits	Voltage Level (Amplitude)
00	+3
01	+1
10	−1
11	−3

Spectral efficiency is a measure of the number of digital bits that can be encoded in a single cycle of a wave form. The duration of a single cycle of a wave form is called the *symbol time*. 2B1Q has the advantage of being a well-understood modulation scheme, relatively inexpensive, and robust in the kind of interference observed in a telephone plant. If, however, you want more b/Hz, other modulation techniques that provide more voltage levels are needed to yield a higher spectral efficiency. To encode 3 bits per symbol time, for example, you need eight voltage levels. To encode k bits in the same symbol time, you need 2^k voltage levels. As speed requirements increase, it becomes increasingly difficult for the receiver to discriminate among many voltage levels with consistent precision.

The limits on spectral efficiency of modulation techniques using amplitude alone preclude its use with very high bit rates, including RBB applications such as video or high-speed data retrieval. More aggressive modulation techniques exist that permit greater spectral efficiency. Among the most important is *Quadrature Amplitude Modulation (QAM)*.

Quadrature Amplitude Modulation (QAM)

Another aspect of the wave form, phase, can be modulated to encode information. The simplest form of phase modulation that does not use any amplitude modulation is *phase shift keying (PSK)*. PSK is not commonly used because of its low spectral efficiency.

A technique that modulates both amplitude and phase is *Quadrature Amplitude Modulation (QAM)*. By using both amplitude and phase modulation simultaneously on a single carrier, QAM yields a higher spectral efficiency than 2B1Q. The phase of a single RF carrier is shifted by 0, 90, 180, or 270 degrees from a reference carrier. It is possible to modulate at 45 degrees or 22.5 degrees, but receivers have more difficulty discriminating among the smaller phase angles.

The number of levels of amplitude and the number of phase angles is a function of line quality. Cleaner lines (with flat frequency response and high carrier/noise ratio) permit more aggressive modulation, which translates into more spectral efficiency (b/Hz).

Various flavors of QAM are referred to as QAM *nn*, where *nn* is an integer indicating the number of states per hertz. The number of bits per symbol time is *k*, where 2^k=nn. If, for example, 4 b/Hz are encoded, the result is QAM 16; 6 b/Hz produces QAM 64.

Quadrature Phase Shift Keying (QPSK) is a particular case of QAM modulation and is equivalent to QAM 4. QPSK achieves a modest spectral efficiency of 2 bits per symbol time, but it can operate in the harsh environments of over-the-air transmission and cable TV return paths. Because of its robustness and relatively low complexity, QPSK is widely used in cases such as direct broadcast satellites.

QAM-16 is proposed as an optional modulation scheme for return-path data traffic on cable plants. The *Society of Cable Television Engineers (SCTE)* has specified the 64 and 256 QAM modulation for digital video transmission over cable TV plants.

Carrierless Amplitude Modulation/Phase Modulation (CAP)

Another variation of amplitude and phase modulation is *carrierless amplitude modulation/phase modulation (CAP)*. As mentioned previously in the case of phase modulation, a reference carrier was shifted some number of degrees to encode information. A CAP receiver is clever in that it is able to derive the carrier, whereas a QAM receiver must have the carrier sent explicitly from the transmitter. CAP and QAM are so similar that, for some implementations, CAP receivers can receive a QAM transmission.

CAP is utilized as a modulation scheme for ADSL.

Multicarrier Techniques

2B1Q, QPSK, QAM, and CAP are examples of well-understood single-carrier techniques with a lot of industrial and defense experience behind them. Satellites, for example, have used QPSK for years.

With the development of *digital signal processing (DSP)*, multi-carrier techniques are now possible. Multicarrier techniques use an aggregate amount of bandwidth and divide it into subbands, thereby yielding multiple, parallel, narrower channels. Each subband is encoded using a single-carrier technique, such as QAM, and bit streams from the subbands are bonded together at the receiver. Important examples of multicarrier techniques are *orthogonal frequency division multiplexing (OFDM)* and *discrete multitone (DMT)*.

Figure 2–2 shows an example of multicarrier modulation using the current ANSI T1.E1 proposal for ADSL. In this figure, 1 MHz is segmented into 256 subbands of 4 KHz. Each subband is modulated by the transmitter using a single-carrier modulation technique. The receiver accepts the subband and bonds the 256 carriers together.

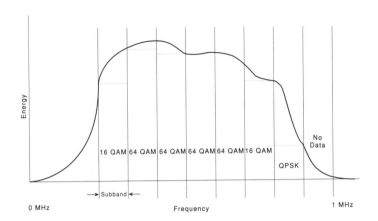

Figure 2–2
Multicarrier modulation.

OFDM and DMT differ in that OFDM uses a common modulation scheme for each subband. That is, each subband transfers the same number of bits per second. OFDM is used in European over-the-air broadcast digital television. In the case of over-the-air broadcast, all subbands are presumed to have uniform noise characteristics, so a common modulation technique makes sense.

DMT enhances the OFDM model by allowing variable spectral efficiency among the subbands. Some subbands can use more aggressive modulation schemes than other subbands. DMT is used in wired media such as xDSL, where the noise characteristics of each wire might differ. Therefore, spectral efficiency can be optimized for each individual wire.

Multicarrier techniques have a latency penalty (time delay to transmit a digital bit) compared with single-carrier techniques. In the DMT case for ADSL, there are 256 subbands of 4 KHz each. This means that no bit can travel faster than allowed by 4 KHz, even if the line were perfectly clean.

One of the noisiest debates about modulation techniques occurs between proponents of DMT and proponents of CAP for use by a new service called ADSL, which is discussed in Chapter 4, "xDSL Access Network." DMT for ADSL uses 256 subbands, whereas CAP uses a single carrier. At the time of this writing, CAP has an advantage over DMT in that it consumes less power (thereby generating less heat) and costs less due to lower complexity and more units in the field. It is easy to see how DMT scales; in fact, DMT has been selected by ANSI T1E1.4. Furthermore, a number of U.S. telephone companies have selected DMT in the largest purchase order of that technology to date nationwide. For the moment, however, CAP enjoys some advantages, and its proponents are sticking to their guns.

Considerations in Selecting Modulation Techniques

Selection of modulation technique for each Access Network has been highly contentious, largely because there's a lot of money at stake. Standards organizations for cable TV, xDSL, and HDTV have spent years arguing the requirements of modulation, let alone the choice of a specific technique. Although commercial self-interest, academic background, national pride, embedded base, and personal ego play a role, there are also engineering and cost tradeoffs to consider.

The following list summarizes some of the major engineering considerations:

- **Scale.** Will the modulation technique support large systems and fast bit rates?

- **Noise immunity.** Can the modulation scheme operate reliably with real-world impairments?

- **Packaging.** Can *Application-Specific Integrated Circuits (ASICs)* be built? Can implementations be used in a variety of environments, such as different Access Networks? How large are the components, and how much power does the technique consume?

- **Performance.** What is the spectral efficiency? What is the latency?

- **Cost.** Naturally, cost is the dominant factor when dealing with consumer markets.

The modulation schemes described in this chapter are likely to be RBB alternatives. Table 2–2 lists services and their respective modulation schemes, current as of this writing.

Table 2–2   *Modulation Techniques for Current Services*

Service	Modulation Techniques
ISDN (USA)	2B1Q
U.S. Direct Broadcast Satellite	QPSK
U.S. Digital Over-the-Air Broadcast	Vestigial SideBand (VSB), (a technique similar to QAM)
U.S. Digital Cable Forward Channels	QAM 64, QAM 256
U.S. Digital Cable Return Channels	QPSK, QAM 16
European Digital Over-the-Air Broadcast	OFDM
European Digital Cable Forward Channels	QAM 64
High bit rate Digital Subscriber Line (HDSL)	2B1Q
Asymmetric Digital Subscriber Line (ADSL)	DMT, CAP

Interoperability of Modulation Techniques

The proliferation of modulation techniques raises interoperability problems for digital television. The likelihood now exists that a television built for over-the-air digital broadcasts will not be able to receive a cable TV digital transmission without a separate box. Otherwise, HDTVs will be built with dual-mode VSB/QAM demodulators.

The consumer could end up with two set tops: a VSB MPEG decoder for digital over-the-air reception and a QAM MPEG decoder for digital cable reception. Both decoders will have analog descrambling built in, as well. There may be new generations of TVs with input jacks for all three types of reception. The results are market confusion and additional costs.

NOISE MITIGATION TECHNIQUES

The signal-carrying capabilities of wired networks will be stretched to the limit by high-speed services, and signal errors caused by noise will result. Because broadband networks will be used to transmit video, errors cannot be tolerated—at least not many of them—without severe picture degradation and even total loss of picture. Errors in data transmission generally are less severe than errors in video transmission because networking protocols can make adjustments. In the extreme, however, errors in data transmission can lead to a loss of session.

Therefore, noise mitigation techniques particularly tailored to fast networking are to be used for RBB. The two most prominent techniques are concatenated Reed-Solomon *forward error correction (FEC)* and *direct sequence spread spectrum*. Forward error correction enables the receiver to detect and fix errors to data packets without the need for the transmitter to retransmit packets. Spread spectrum enables the successful transmission in hostile transmission environments.

Forward Error Correction (FEC)

Forward error correction (FEC) techniques such as Reed Solomon (RS) are vital for the system to maintain speed. Forward error correction enables the receiver to detect and fix errors to data packets without requiring the transmitter to retransmit packets. If retransmission techniques were used, a great deal of latency would be incurred by the round-trip of negative acknowledgment of the original packet and retransmission of the second packet. Three trips would have to be made across the network. In other systems, such as broadcast television, there is no back channel to accomodate retransmission techniques.

The technique works when errored bits are spread out, but if too many errors are grouped together, the technique fails to identify the errored bit.

Interleaving

Interleaving, within the context of error correction, is a process whereby bits are reordered so that errors due to impulse noise are spread over time. The top of Figure 2–3, for example, shows a transmission sequence of 24 bits, with a burst error affecting bits 7–10. In the interleave process, the bits are reordered into an interleave buffer, as shown in the figure. The errored bits are now spread out, and forward error correction is applied to the reordered bits in the interleave buffer. This illustration shows a simple reordering in the interleave buffer, but in practice, more sophisticated reordering schemes are used. The reordering, for example, might differ from block to block. This creates robustness in the presence of noise characteristics that are repetitive.

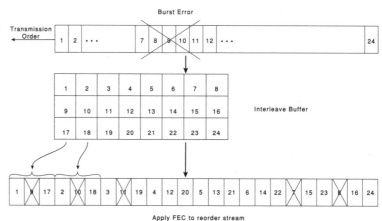

Figure 2–3
Interleaving in forward
error correction.

By concatenating interleaving with forward error correction, errored bits can be reduced by factors of up to 10,000. Results presented in the ATM Forum (ATM Forum 95-1154; Esaki, Guha) indicate an improvement in packet error rate of 10^5 for full-sized AAL5 (65K byte) packets and 10^2 for 9180 byte packets.

The price of interleaving is latency. The entire bit stream experiences delay because the bits must be taken in and out of the interleave buffer (which results in serialization delay) and because Reed-Solomon calculations must be performed. The amount of latency increases with the number of bits interleaved, and more burst protection means more latency. Experts have argued that it is possible that a transmitter can transmit at near the theoretical limits of a medium by extending the interleaving. Because there is a necessary tradeoff of signal quality and latency, compromises must be made.

DAVIC has settled on an interleaving latency that protects against burst impulse noise of 25 microseconds, yielding latency of 600 microseconds in duration.

But American interests (particularly CableLabs) argued for a more flexible scheme than fixed interleaving. As a result, the Multimedia Cable Network System (`http://www.cablemodem.com`) in the United States has specified a variable depth interleaver that protects against burst-impulse noise from zero to 96 microseconds. The International Telecommunications Union (ITU) is in the process of specifying modulation and forward error control standards for digital TV. As part of its deliberations, two options exist for interleaving specified in annexes A and B of the digital TV specification. These two options are described in Table 2–3.

Table 2–3 *Comparison of ITU—Annex A and Annex B*

	Annex A	Annex B
Latency	600 micro	0–3,750 microseconds
Maximum Burst	25.6 micro	0–96.0 microseconds
Reed Solomon	(204,188)	(128,122)
Supporters	DVB IEEE 802.14 DAVIC	MCNS Partnership (see Chapter 3) IEEE 802.14

Web references to each of these supporting organizations are presented in Appendix A, "RBB-Related Companies and Organizations."

Notation

Typical Reed-Solomon approaches to FEC involve interleaving and coding. The notation for Reed Solomon is characterized by (N,K), where N is the total number of bytes per block and K is the intended number of data bytes per block. The quantity $(N-K)/2$ is the number of errored bytes per block correctable by the coder.

For example, RS(204,188) denotes the shorthand notation for RS encoding for MPEG packets. An MPEG packet is 188 B in length, 16 B are added for error correction, yielding a block size of 204 B. The burst error protection is 8 B per block (16 divided by 2).

FIBER-OPTIC TRANSMISSION

Fiber-optic technology is a major reason for the feasibility of RBB. More than metal wires, fiber has the carrying capacity to supply high-speed services to millions of homes. Fiber-optic cables are made from clear glass—much more transparent than window glass. Fiber technology also uses lightwave signals to transmit data, whereas metal wire uses electromagnetic signals. The clarity of the glass permits the distribution of lightwave impulses for possibly hundreds of miles.

Coding is simple. A transmitter indicates a binary 1 by higher intensity light, if light is lower intensity it indicates binary 0. Light is perceptible when the receptor senses photons. Of course, more complicated coding schemes are possible, but a simple on/off delineation works with current technologies.

Description of Key Fiber Elements

This section examines three central aspects of the fiber system: the fiber itself, the light sources, and modulation techniques.

Fiber

Fibers are broadly categorized into two groups; single-mode fiber and multimode fiber. The term *mode* refers to the path a photon takes in going from one end of the fiber to another. In multimode fiber, a photon careens off the fiber wall as it goes from one end to the other, thereby defining a path. Another photon (there are a lot of them) probably will take a different path. The number of possible paths is a function of the core diameter and is on the order of 200.

According to the recommendations in ANSI T1E1.2/93-020R3, the core of multimode fiber is 62.6 microns in diameter, and the cladding is 125 microns. This is compatible with the FDDI multimode fiber specification ISO/IEC 9314-3.

As light bounces off the fiber core, it loses some energy. This limits the distance over which multimode fiber is usable without amplification. A second effect is that because the paths of photons that make up a data bit are of different lengths, a given bit is spread in time. Figure 2–5 shows two photons following different paths in a multimode fiber. Photon B will arrive at the destination before Photon A. This time difference between the arrival of Photon A and Photon B is called the *delay spread*. Because Photons A and B are part of the same symbol (binary bit), significant delay spread can cause interference between bits. This is called *intersymbol interference (ISI)* and is an impairment for metallic and wireless communications.

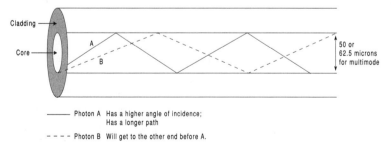

Figure 2–4
Fiber cross-section.

For single-mode fiber (sometimes called *monomode*), all photons take the same path down the center of the core. This is because the core of the fiber is very narrow. According to the recommendations in ANSI T1E1.2/93-020R3, the fiber has an 8.3 micron core with 125 micron cladding. Because much less signal loss and less ISI results with single-mode fiber, this type is capable of greater distances and higher bit rate. The fact that the diameter of the fiber-optic core is narrower than that of multimode fiber means that other elements of the system—such as connectors, transmitters, and receivers—must operate with much smaller tolerances. This makes the other components more expensive and difficult to handle.

A variant of fiber-optic cable is *Plastic Optical Fiber (PoF)*. PoF is made of clear plastic, which is not as transparent as glass and thus can be used only for short distances. PoF, however, is less expensive and more malleable than glass. These features make it interesting for indoor use. PoF is discussed further in Chapter 7, "In-Home Networks," in the context of Home Networks.

Light Sources

Fiber can be illuminated either with lasers or *light-emitting diodes (LEDs)*.

Lasers are used to illuminate fiber over long distances. A light source is injected into one end of a fiber network at a certain launch power (brightness). The light through the medium is of the same frequency, yielding a coherent beam. This means that all photons take the same time to traverse the fiber.

Lasers can be of three major types. In increasing order of cost, these types are *Fabry Perot, Distributed Feedback (DFB),* and *Vertical Cavity Surface Emitting Lasers (VCSEL)*. Fabry Perot lasers have less signal-carrying capacity than DFBs, which in turn have less performance than VCSELs. *Erbium Doped Fiber Amplifiers (EDFA)* amplify optical signals for very long distance service. With EDFA, laser signals can cross the United States without electro-optical repeaters. Costs for Fabry-Perot lasers are less than $1,000, while costs for DFB lasers can be up to $12,000, depending on launch power. Costs for EFDAs are higher than DFB.

Where bit rate requirements are less than a Gb and cost constraints are severe, it is possible to use light-emitting diodes (LED). LEDs cost only a tiny fraction of the cost of a laser and consume far less power. Light emitted from LEDs, however, is not coherent, so multiple frequencies exist. Hence, potential for ISI exists, which lowers bit rate. LEDs can be used for up to 200 Mb, which is suitable for in-home use but not for carrier use. PoF is illuminated with LEDs.

Modulation Techniques

The fiber can be modulated using on-off keying, frequency modulation, or amplitude modulation. Symbol times are very short, on the order of picoseconds. The computer industry tends to think of fiber as a digitally encoded medium (which it can be), but the cable industry requires analog line coding.

When cable TV was implemented, it was intended to mimic over-the-air analog broadcast signals. This meant that cable signals were analog as well. If fiber was being encoded digitally, then it would have been necessary to convert from analog to digital (for fiber transport) and back to analog (for display on analog TV sets). The cable industry wanted to avoid extra analog to digital (A/D) conversions, even though this process would produce possible benefits of data compression and improved picture fidelity. A/D conversions were just too costly. Amplitude-modulated fiber requires the use of linear lasers, which are significantly more expensive than digital lasers designed for on-off keying. Amplitude-modulated fiber enabled analog line coding, which was the key that propelled the cable industry to deploy fiber.

Benefits

As compared with metal, fiber has more bandwidth, can travel longer distances without amplification or regeneration, is safer to handle, weighs less, and has much lower material cost. Single-mode fiber can carry light for up to 70 km, whereas coaxial cable must be amplified every 0.5 km. Phone wire has a reach of maybe 5 km, or up to 18,000 feet.

Fiber does not rust, which translates into less outside plant maintenance and lower ongoing operational costs for local loop networks. Fiber also is impervious to electromagnetic interference and is secure. Furthermore, it is extremely difficult (though not impossible) to wiretap fiber, which is either good or bad, depending on which side of the law you stand.

Fiber also promotes high reliability. A service break caused by a fiber cut can be corrected by switching to a redundant network in approximately 50 ms. This nearly instantaneous switch won't be noticed by consumers or most networking applications. Carriers are moving to ring and other redundant topologies to secure their fiber networks.

Impairments

Fiber is wonderful stuff, but it has its limitations. As with metal wire, fiber is subject to attenuation in certain circumstances. Plus, other impairments—such as attenuation, dispersion, handling problems, cuts, bends, and clips—are unique to fiber.

Attenuation

In fiber, *attenuation* refers to the loss of optical power as light travels through the fiber. Measured in decibels per kilometer, attenuation ranges from over 300 dB/km for plastic fibers to around 0.2 dB/km for 1,550 nm for single-mode fiber (a nanometer is a billionth of a meter). For 1,310 nm of fiber, loss is about 50 percent greater than for 1,550 nm. Additionally, each fiber splice adds attenuation of about 1 dB per mechanical splice. A good fusion splice can be roughly 0.1 dB. Due to fiber age, chemical makeup, construction, installation care, and other factors, the fiber loss per kilometer might not be the same for every link.

Also, cracks occur in the fibers over time, sometimes caused by stresses introduced during the installation process. In freezing conditions, fiber-optic cable can be damaged when moisture inside imperfect insulation turns to ice. As ice crystallizes, it can exert crushing pressure on the fiber cable inside the conduit. That pressure can cause fissures in the fiber and thereby degrade the signal, especially as higher speeds are reached. The pressure also could break the fiber and kill the signal altogether.

Dispersion

Dispersion causes light to spread as it travels down a fiber and therefore limits bandwidth. Some photons travel straight down the middle of the fiber; other photons bounce off the walls of the fiber and go careening through the fiber. The photons going down the middle will arrive earlier than the bouncing photons, so the effect of dispersion is the loss of temporal integrity. The bit rate through the fiber must be low enough to ensure that pulses do not overlap.

The question arises whether attenuation or dispersion is the more significant limiting factor in fiber transmission. The answer depends on speed, attenuation, and optical loss budget. Most cable operators operate with a maximum optical budget of around 37 dB. Because 1,310 nm of single-mode fiber attenuates at 0.35–0.5 dB per kilometer (with fiber attenuation, splices, bends, kinks, and a safety margin), a distance of 74 km can be supported. For OC12 over 1,310 nm single-mode fiber, the dispersion distance is about 40 km. For OC3, which is slower but more forgiving than OC-12, distance is about 160 km.

Handling Problems

Fiber is difficult to handle because electrons are more forgiving than photons. The core of single-mode fiber is nine microns, which makes splicing very difficult and also could impair the construction of connectors and sockets. The splices must be scrupulously clean and absolutely flush, and the ends of the splices must be perfectly aligned. No alignment problems occur with electricity—any little contact, and electrons move. This is not so with fiber.

Handling problems become particularly acute for Fiber to the Home. Great care and fine tools are needed to align the glass cores on the respective segments. New devices are available, which do splice automatically.

Cuts and Other Damage

Fibers can be cut more easily than copper, and when they are cut, the effects are large. One industry rule-of-thumb points toward one fiber cut/year/500 km of fiber. Alcatel says that typical fiber breaks require 6–12 hours to locate and repair. Separately, *Fiber Optics News* reports 390 fiber cuts in the United States in the year ending June 1996. Reported fiber outages not caused by cuts—including storms, fire, vandalism, and rodent damage—were 247. The catastrophic effect of cuts means that redundancy is a must, thereby greatly increasing the number of fiber miles of a system.

Bending

If copper wire is bent, electricity flows through the bend without loss. If fiber is bent excessively, though, light escapes the core and the fiber might break. The amount of tolerable bend is a function of the refractive index of the core and the surrounding cladding. Fibers specify a minimum bend radius; any tighter bending than the specified radius can stop the transmission altogether. Most single-mode fiber can tolerate a bend radius no less than 3 cm. Figure 2–5 contrasts acceptable and unacceptable bend radii.

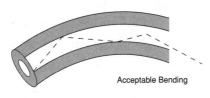

Acceptable Bending

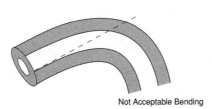

Not Acceptable Bending

Figure 2–5
Bending limits.

The ATM Forum is considering using plastic optical fibers for home networking. This would provide very high-speed networking in the home. Working around the nooks and crannies of a home might, however, require lots of bending, rendering fiber or plastic unusable.

Clipping

Fibers transmit signals within an allowable dynamic range that cannot be subjected to too much or too little laser power. When excessive power is input into an amplitude modulated laser, the receiver may reach its maximum power level. This is known as laser *clipping*. Clipping can occur when several sources that have common phase and amplitude alignment are input into a single laser transmitter simultaneously. Clipping primarily affects wide-area transmission and is a potential problem for hybrid fiber coaxial cable systems.

An Improvement to Fiber

Even with the continued purchase and installation of fiber, more capacity will be needed. Fiber is expensive to install. In fact, underground fiber installation in urban areas can be as high as $70,000 per mile. This means that carriers are looking for ways to better utilize the fiber they have.

Dense wavelength division multiplexing (DWDM) is a new technology that provides in fiber the equivalent of frequency division multiplexing in metallic wire. The idea is that separate parallel channels are transmitted on a single fiber, with one wavelength (or color) for each channel. The color differences among channels on a DWDM system all appear red to the human eye but are easily seperated by the system. Current products enable 16 channels of 2.5 Gb each for a total of 40 Gb per fiber. DWDM can operate over existing single-mode fiber and therefore substantially reduces upgrade costs for large carriers.

Fiber-Optic Market

Major improvements in fiber and laser technology are expected to further stimulate the deployment of fiber. The following list discusses some key projections:

- Kilometers of fiber-optic cable worldwide (1995): 22.8 million; Expected in 2001: 63.2 million

- Kilometers of single-mode fiber installed in the United States (1996): 10 million; Expected in 2001: 16 million

- Bit rate of most current single-mode fiber: 2.5 Gbps; Bit rate of a single fiber using DWDM: 40 Gbps

- Current distance between electro-optical amplifiers: 360 km; Distances expected with EFDA: 2,000 km

The growth of RBB and the growth of fiber are linked. Fiber has the carrying capacity to move data across Transport Networks, so it will be key to scaling some Access Networks, particularly digital subscriber loop networks. Fiber might even find its way into the home.

SIGNALING

Signaling refers to the process by which an end system notifies a network that it wants service. The network responds with resources (bandwidth, buffering, and entries in databases) that enable the connection to proceed. If these resources are not available, the network notifies the end system. Signaling is used to release resources at the end of a session, and it is used during a session to change attributes. Signaling, for example, can add bandwidth, a third party, or other capabilities such as access to services.

The most familiar example of signaling is picking up a telephone and dialing a phone number. A computer at the telephone company notices when you pick up the phone. Upon receiving the phone number, a route is established to your requested destination, and bandwidth is reserved. If you dial a 500, 800, 888, or 900 number, there is a database lookup to find the "real" phone number that is hidden from the caller, and a route is established to the phone number found in the database. Recent innovations, such as the implementation of a packet network in the telephone network, enable all this signaling to occur more quickly than previously. Even so, the call-setup process is time-consuming and expensive. Modern switches are capable of only a few hundred call setups per second.

Signaling requires a round-trip from the origination point of the network to the destination to determine whether sufficient resources are available at that instant in time. A round-trip moves through the Access Network and the Transport Network, during which lots of events take place. Switches reserve resources (such as memory), databases are consulted, transmission links are checked for bandwidth availability, and end systems are validated for proper addressing.

As an alternative to the significant housekeeping chores of telephone signaling, connectionless networking emerged as a highly scaleable networking architecture. The Internet is the paradigm of connectionless networking.

Connectionless protocols such as IP impose a minimal signaling burden on data communications equipment and end-user devices because the protocols do not have a call-setup process. If resources are available, the packet arrives. If not, the packet is discarded, and no call setup takes place. It's easier to ask for forgiveness than for permission.

Signaling protocols exist for all networks. In addition to the telephone example cited previously, examples of signaling protocols are the *Resource Reservation Protocol (RSVP)* for IP networks such as the Internet; ATM Q.2931, which establishes call setups for ATM; and various flow-setup protocols for packet switching.

VIEWPOINT

Signaling Rate as an RBB Bottleneck

When one hears of RBB networks, one generally thinks of the requirement for bandwidth and speed. A strong argument, however, can be made that the fundamental technical bottleneck for RBB is not bandwidth but *signaling rate*. To use an extreme example as an illustration, if a telephone switch were capable of only one call setup per minute, then it wouldn't matter that the switch can transmit 1 Gbps of information. The throughput would never be realized because so few callers could use the switch (of course, no one can talk that fast anyway).

The same holds true for data-switching equipment. If routers and ATM switches cannot support sufficiently high signaling rates, then the bandwidth offered by fiber-optic networks cannot be realized.

Current measurements on the Internet show that one can expect on average less than 20 packets per data flow and that traffic is highly bimodal in terms of packet length. That is, packets are either very short (less than 64 bytes) or very long (1,500 bytes—about the maximum length of an IP packet). A large fraction of Internet packets are acknowledgments, which are of minimal length.

The impact of short flow length is to increase the signaling required to sustain maximum throughput through the network. If, for example, an IP switch has 16 ports of OC-48 (2.488 Gb) port speed each, aggregate speed of the switch is 40 Gb. If average packet size is 256 B, then the router must transmit 20 million packets per second. If there are 20 packets per flow, then the router must be capable of 1 million flow setups per second. This is far beyond the capability of any IP switch, ATM switch, or voice switch available today, or likely to be available in the near future. These statistics suggest that signaling rate will choke the switch long before packet forwarding is choked.

These calculations are not meant to suggest that a million flow setups per second are needed for RBB. In fact, it is not known what the required flow setup rate is, because the applications are not built yet. The point is that the bottleneck through a network might be signaling, not bandwidth. Therefore, applications and networks should be designed with signaling rate in mind.

IP MULTICAST

IP multicast is what makes push-mode data work over point-to-point networks. In particular, IP multicast reduces signaling, reduces data volume, and tends to synchronize reception of common data flows to multiple receivers. These are useful and even necessary characteristics for mass market network services.

In the hybrid fiber coax and wireless cases, broadcasting of data is a natural consequence of the physical medium. In the point-to-point wiring case, broadcasting of data is achieved by *multicast*, a software technique that replicates packets efficiently. Multicast is more difficult to implement than broadcast because it must track who will and who won't receive packets. Broadcast assumes that everyone within receiving range should receive the packet.

Figure 2–6 depicts the unicast case in which three clients exist on a network. Rosie and Jimmy are supposed to receive packets for a particular application from the server, and Junior is not. In this case, the server replicates packets for Rosie and Jimmy, and the router R1 forwards both packets.

Figure 2–6
Unicast replication of
packets.

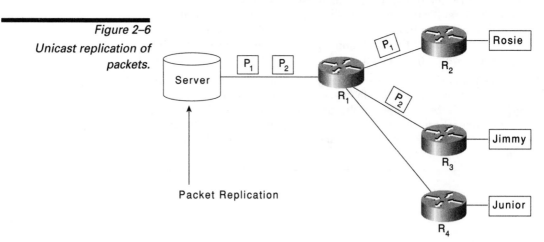

Packet Replication

Figure 2–7 depicts the multicast case for the same application. In this case, the server emits one packet, and the router R1 replicates the packet for Rosie and Jimmy. Work is distributed from the server (a single point) to the network, where many routers can replicate packets. The link between the server and R1 carries only one packet, versus two for the unicast case. Note that R1 is smart enough not to replicate a packet for Junior. Multicasting requires considerably more intelligence in the routers than either the unicast or broadcast models. One reason is the requirement to build packet-distribution trees, which list the nodes through which packets travel to reach all multicast receivers. In Figure 2–7, the distribution tree consists of the server (which serves as the root of the tree) and routers R1, R2, and R3, but not R4.

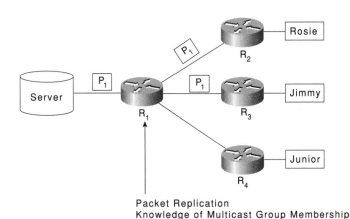

Figure 2–7
Multicast replication of packets.

A primary goal of these packet-distribution trees is to identify locations where packet replication is performed and thereby minimize replication. That is, if multiple receivers exist on a given branch of the network, only one copy of the packet should be sent to that branch, where it then can be replicated by an appropriate router.

To take a more complicated scenario, the server can provide broadcast video services to possibly hundreds of viewers. The server is connected to 10 routers, each of which is connected to 10 routers, which in turn are connected to 10 more, and so on. Figure 2–8 illustrates this example. Imagine that the total population is 100,000 viewers, of whom 1,000 are viewing a program at a particular moment.

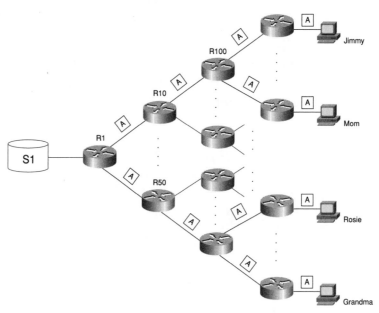

Figure 2–8
A more complex multicast.

It would be infeasible for the server to create 1,000 copies of every packet to go to each viewer. In this case, the server creates only five packets, and other routers downstream from the server replicate as needed so that all 1,000 viewers get the bit stream. As long as only one server or source of video exists, this is a system that can scale to ultimately serve perhaps millions of viewers.

Multicast is a crucial technology for scaling for RBB networks because it enables a sender and the point-to-point network delivering the content to transmit the minimal amount of data to the consumer with a minimal amount of signaling. Because Apple-Talk and IP are the only networking protocols supporting multicast, they are the only two data-networking protocols that can be considered for RBB use.

End System Requirements

Registration is the mechanism for the client to inform the network that it is a member of a particular multicast group. RFC 1112 defines the *Internet Group Membership Protocol (IGMP)*, which specifies how hosts inform the IP network that it is a member of a particular multicast group. (RFC is a Request for Comment, which specifies an Internet recommendation. A comprehensive list of all Internet RFCs is located at http://www.isi.edu; click the Services icon.)

Additional management software might be needed so that the multicast is open only to those who are authorized to join the multicast group (for security reasons) and identifies the duration over which an end station has participated in the multicast (for tariffing reasons).

By joining a multicast group, the host performs the equivalent of tuning to a channel in the broadcast analog world. By tuning into a multicast group, the receiver subsequently can passively receive data flows without incurring the network and server overhead of establishing a new connection and with minimal packet replication.

IP Multicast Routing Protocols

This section discusses the following two standards available for routing IP multicast traffic:

- Distance Vector Multicast Routing Protocol (DVMRP), (RFC 1075)

- Protocol-Independent Multicast (PIM), operating in sparse mode and dense mode

Apple's Simple Multicast Routing Protocol (SMRP) is used for multicast routing of AppleTalk.

Distance Vector Multicast Routing Protocol (DVMRP)

DVMRP (RFC 1075) is the earliest of the IP multicast routing protocols. It has supported the Internet *MBONE (Multimedia Backbone Network)* for years. DVMRP uses a technique known as *reverse path forwarding*. When a router receives a packet, it floods the packet out of all paths except the path on which it received the packet. This is the same action that occurs in the broadcast case, in that packets are sent to routers downstream that do not want them. When a router receives a packet destined for a multicast group for which it has no receivers, it returns a "prune" message back to the sending router, telling the sending router to stop sending packets for that multicast group.

DVMRP periodically refloods to reach any new hosts that want to receive a particular group. A direct relationship exists between the time it takes for a new receiver to get the data stream and the frequency of flooding.

DVMRP implements a proprietary unicast routing protocol in the return path route to "prune" messages back. This unicast routing protocol is based purely on *hop counts*, that is, the number of routers traversed. As a result, the forward multicast traffic might not take the same path that the "prune" messages take.

DVMRP has significant scaling problems because of the necessity to flood frequently. In some instances, the source floods to find new clients when there are none. This creates lots of unnecessary traffic.

Protocol-Independent Multicast (PIM)

Protocol-Independent Multicast (PIM) supports two different modes of traffic patterns: dense mode and sparse mode.

Dense mode is most useful under the following circumstances:

- Senders and receivers are in proximity to one another.

- There are few senders and many receivers.

- The volume of multicast traffic is high.

- The stream of multicast traffic is constant.

Dense-mode PIM uses reverse path forwarding and looks a lot like DVMRP. The most significant difference between DVMRP and dense-mode PIM is that DVMRP requires its own RIP-like routing protocol for returning the "prune" messages. PIM does not require any particular unicast protocol, hence the term *protocol-independent*.

Sparse multicast is most useful under the following circumstances:

- There are few receivers in a group, or the receivers are widely dispersed.

- Senders and receivers are separated by WAN links.

- Traffic is intermittent.

These are characteristics of RBB, so PIM-sparse likely will be the multicast protocol of choice for RBB.

Sparse-mode PIM is optimized for environments in which many multipoint data streams exist. Each data stream goes to a relatively small number of users. For these types of groups, reverse path forwarding techniques waste bandwidth.

Instead of "flood and prune," sparse-mode PIM works by designating specific routers as rendezvous points, where senders register their flows and receivers register requests to receive flows. When a sender wants to send data, it first sends to the rendezvous point. When a receiver wants to receive data, it registers with the rendezvous point. After the data stream begins from the sender to the rendezvous point to the receiver, the routers in the path optimize the path dynamically to remove any unnecessary hops. Sparse-mode PIM assumes that no hosts want the multicast traffic unless they specifically ask for it.

Benefits of IP Multicast

It is believed (it isn't known for sure yet because no one has built a large consumer multicast network) that multicast flows tend to be longer in time duration, and that packets will, on average, be larger than flows and packets in a unicast model. Longer flows occur because they are not interrupted by user-initiated signaling as frequently. Longer packets exist, on average, because the server can package data into larger packets and because there will likely be fewer acknowledgments that are of minimal packet size. Longer flows and longer packets tend to increase bandwidth and decrease switch utilization.

Finally, there is conjecture that multicast is more compatible with an advertising model for data distribution than HTTP. In an advertising model, content is controlled at a central source for wide distribution with minimal consumer interaction. The advertiser does not rely on the client to click an icon to receive a screen full of advertising; the advertising simply is sent.

Challenges of Multicast

Multicast is complicated software that resides in routers. For it to be deployed, Internet service providers (ISP) must put multicast in their routers. Before ISPs will be willing to do so, accounting and security issues must be addressed.

Accounting

Charging for participation in multicast groups will stimulate ISPs to provide more multicast services. The problem is how to charge for the service. Tariffing can be enforced by duration of the multicast, the speed at which the receiver can accept packets, and membership fees. To apply tariffs, the service provider must know who is on the multicast at any time, when the users leave, at what quality of service the users connect, and the total duration of the session. Because a single source links with clients in multiple geographic areas, it is unlikely that multicast tariffing will have any distance sensitivity. Such measurements are works in progress as of this writing.

Security

Content providers of multicasts might have restrictions on who can participate and whether traffic is encrypted. Along with the requirement to tariff the service, some form of directory service probably is required. Variations of security are detailed in the following list:

- **Unrestricted.** Anybody can come and go as desired, and traffic is not encrypted.

- **Restricted and not encrypted.** A limited list of subscribers can come and go as desired, but the traffic is not secured via encryption.

- **Restricted, encrypted.** A limited list of subscribers can come and go as desired, and traffic is encrypted.

An encrypted multicast group will have a unique key, which will be distributed securely to users while the multicast is in session.

Scaling

Figure 2–8 illustrates the feasibility of multicast serving thousands or millions of users. This figure shows a scenario in which only one program is served to users on the network.

A scaling problem arises when multiple programs exist that traverse a common set of routers. In such a case, the routers need to prune and replicate for multiple packet-distribution trees.

Figure 2–9 shows two servers—S1 and S2—that broadcast to the same population by using the same set of routers. Router R50 replicated packets only for S1 in Figure 2–9. With the addition of server S2, the same router must replicate packets for the original multicast and packets for the second multicast. In the process, S2 computes two routing protocols, one for each packet on the distribution tree. Mom, for example, might disconnect from the S1 multicast and decide to join the S2 multicast. In this case, router R201 must be added to the S2 multicast group and must be pruned from the S1 multicast so that it forwards S2 packets instead of S1 packets. Router R201 would receive the S2 packets from router R101. Further research is needed to understand the limits of multicast as a function of the number of multicast groups.

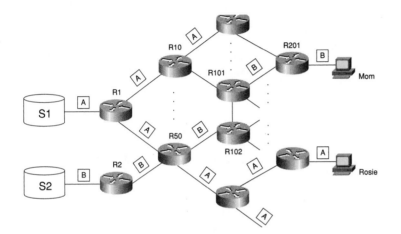

Figure 2–9
Multicast scaling.

THE LIMITS OF AUDIO/VISUAL PERCEPTION

Although networks are limited by what they can deliver economically, what humans can perceive is also limited. Human perception is not infinitely sensitive, so some coarseness of data can be accepted, which tends to make life easier for the network developer.

The limit of human auditory sensitivity, for example, lies in the range of 20 Hz to 20 KHz per second. Obviously, a network providing audio service above that pitch is pointless.

Film is shown at 24 frames per second. Some film is hand-crafted to eliminate motion artifacts at 24 frames per second. Therefore, generally showing at higher rates than this is not needed, because viewers can't tell the difference anyway. One exception to this rule, though is that humans can tell a difference with flashing lights. Flashing lights at 30 frames per second (fps) is noticeable, for example.

Other important picture rates are detailed in the following list:

- U.S television: 30 fps

- European television: 50 fps

- Computer monitors: 72 fps

Human motor response for interactive games is less well calibrated because some people are quicker than others. If human motor response is less than 1/75th of a second (about 15 milliseconds), then computer monitors fail to deliver interactive experience for interactive games. If the response is less than 1/30th of a second, then televisions will fail.

Latency for interactive games therefore must be less than 33 ms to produce them on an NTSC monitor, and less than 15 ms on a proscan monitor. This time budget is the sum of two-way transmission and processor delay.

The standard of 100 ms is used by the phone industry as the threshold of human voice response. If voice relay takes longer than 100 ms round trip, then the conversants can't differentiate a speaker's pause from network delays. This is also a useful number to use as an upperbound figure for any latency that requires hand-eye coordination. If you subtract the time it takes the monitor to display an image, then network and computer-processing delays are bound by 67 ms (100 ms–33 ms) for TV sets and 85 ms (100 ms–15 ms) for PC monitors.

AT ISSUE

Bandwidth Guarantees Versus Prioritized Packets

A corollary to the viewpoint that signaling might be the bottleneck in RBB networks is the view that it is preferable to prioritize packets rather than offer bandwidth guarantees in the network to provide reliable video transport.

In software terms, this raises the question of whether it is worthwhile at all to support the Resource Reservation Protocol (RSVP) and ATM Switched Virtual Circuits, or whether to abandon bandwidth guarantees and route individual packets based on a priority setting in the packet.

This is an important question, because bandwidth guarantees require signaling, an end-to-end exchange of information among network elements as to the availability of resources (bandwidth and buffers) at the time the call request is made by the consumer. If resources are not available, the call is blocked, just as the phone system operates.

The end-to-end resource identification takes time, and as indicated previously, this can result in the bottleneck of network processing. So far, the industry supports bandwidth guarantees. At low levels of consumer acceptance, as during the rollout of new RBB services, end-to-end signaling might not be an issue. If, however, transport schemes such as packet or Sonet develop, then the question likely will recur.

MPEG-2 COMPRESSION

MPEG-2 (Moving Picture Experts Group) refers to a family of protocols designed for encoding, compressing, storing, and transmitting audio and video in digital form. MPEG-2 is the protocol of choice for digital video of the HDTV Grand Alliance, but it supports both progressive and interlaced displays.

MPEG-2 is capable of handling standard-definition digital television (SDTV), HDTV, and motion picture films. MPEG-2 supports data transmission, which is used for sending control information to digital set top boxes, and can be used for transmitting user data such as web pages. Furthermore, MPEG-2 is backward-compatible with MPEG-1, which means that MPEG-2 decoders can display MPEG-1-encoded files, such as CD-ROMs. This protocol family also has full functionality for video on demand (VoD). MPEG-2 chips are on the market for real-time encoding, and a specification for MPEG-2 adaptation over ATM AAL5 exists.

MPEG-2 is rapidly assuming a central role in broadband networking, and it is easy to see why. Without digital compression, RBB would not be possible because uncompressed video consumes too much bandwidth (see Table 2–4).

Table 2–4   *Uncompressed Video Bandwidth Requirements*

Format	Pixels per Line	Lines per Frame	Pixels per Frame	Frames per Second	Millions of Pixels per Second	Bits per Pixel	Megabits per Second
SVGA	800	600	480,000	72	34.6	8	276.5
NTSC	640	480	307,200	30	9.2	24	221.2
PAL	580	575	333,500	50	16.7	24	400.2
SECAM	580	575	333,500	50	16.7	24	400.2
HDTV	1920	1,080	2,073,600	30	62.2	24	1,492.8
Film	2000	1,700	3,400,000	24	81.6	32	2,611.2

With numbers such as those in Table 2–4, compression is a must for storage and transmission. For digital broadcast video and film, MPEG-2 is the standard among broadcasters and cable operators.

This discussion of MPEG focuses on two technical characteristics: video compression and systems operations.

MPEG-2 Video Compression

The formal specification for MPEG-2 video is written in ISO/IEC 13838-3, "Generic Coding of Moving Pictures and Associated Audio: Video," April 1995.

MPEG-2 achieves compression in two modes: spatial compression and temporal compression.

Spatial Compression

Spatial compression refers to bit reduction achieved in a single frame. This compression method uses a combination of tiling the image into blocks of bits (8×8), using discrete cosine transformation, and applying Huffman run-length encoding to achieve bit reduction. Spatial encoding is lossy (that is, some information is lost in the compression), but *knobs* (control parameters) exist to trade off image loss against compression and processing. Some of these knobs are user-controlled; others are automatically adjusted in the encoding process.

Temporal Compression

Of greater interest for network analysis is *temporal compression*. Unlike spatial compression, this compression technique achieves the majority of bit reduction. Temporal compression refers to bit reduction that is achieved by removing bits from successive frames by exploiting the fact that relatively few pixels change position in 1/30th of a second (the time period between adjacent frames). A picture can be coded by using knowledge of a prior picture and applying motion prediction so that just the motion vectors of blocks or macroblocks are sent instead of coding a complete picture. This achieves superior compression than, say motion JPEG, which uses the same spatial compression as MPEG-2 but has no provision for temporal compression.

In addition to compression, the quality of video presentation is affected by other factors, which are detailed in the following list:

- Viewing distance

- Bit rate

- Picture content

- Size of display monitor

These factors must be considered when setting knobs for temporal compression. With respect to bit rate, focus group testing [Cermak] has shown that 3 Mb MPEG-2 video is comparable to normal analog TV and VHS over a wide range of content. Increasing bit rates to 8.3 Mb, however, does not improve viewing quality significantly. Instead, 8.3 Mb MPEG was viewed as comparable to the original uncompressed signal.

The following section discusses frame types, which are central to temporal compression.

Frame Types

To achieve temporal compression, the following three different kinds of frames are defined.

- *I-frames*, or intra-frames, which are complete (spatially compressed) frames

- *P-frames*, or predicted frames, which are predicted from I frames or other P frames using motion prediction

- *B-frames*, or bidirectional frames, which are interpolated between I and P frames

P-frames achieve a bit reduction on the order of 50 percent from their corresponding I-frame. B-frames achieve bit reduction on the order of 75 percent. These compression estimates are for 4

Mb MPEG-2 video with SDTV definition and not too much motion. Actual bit reduction differs according to picture content, the mix of I-, P-, and B-frames in the stream, and various knob settings for spatial compression.

Frame Ordering and Buffering

Figure 2–10 shows a sequence of MPEG-2 frames in display order. This order is different from the order in which frames are decoded at the consumer set top.

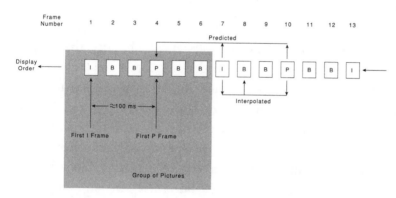

Figure 2–10

MPEG temporal compression in display order.

The decoder receives an I-frame (Frame 1). The next frame received is Frame 4, which is a P-frame. Actually, this is a bit of a misnomer. A P-frame is not a picture sent by the encoder. Rather, it would be more precise to say a P-frame is derived by the decoder from information (motion vectors) sent by the encoder. Frames 1 and 4 are buffered in the decoder and are used to compute B-frames 2 and 3 by interpolation.

The use of B-frames means that the order in which frames are displayed by the receiver is not the same order in which frames are decoded at the receiver. In Figure 2–10, the display order for the first four frames at the receiver is I, B, B, P. The order of decoding, however, is I, P, B, B. This is because the P-frame is computed first and then used with the I-frame to compute the two intervening B-frames.

An I-frame (Frame 7) is transmitted. The original I-frame (Frame 1) is discarded from the decoder buffer memory and the first P-frame (Frame 4) is used with Frame 7 to interpolate for B-frames 5 and 6. With the receipt of the new I-frame, the process is repeated. If many I-frames exist, picture quality tends to be higher, but so does bandwidth use. Still, frequent use of I-frames is necessary when motion compensation fails to produce good pictures.

The use of B-frames is a controversial issue because they impose buffering requirements on the order of 1–2 MB and latency at the receiver. Some vendors—such as General Instruments, an expert and pioneer in the field of digital TV (original Grand Alliance participant)—have resisted the use of B-frames, believing that the cost to the decoder does not justify bandwidth reduction on the network. Attitudes about B-frames are becoming more friendly as the cost of memory drops, but there are still important tradeoffs in buffer size, latency, bandwidth, set-top memory cost, and picture quality.

Using Figure 2–10 again, some control parameters can be defined. First, the term *group of pictures* refers to the set of frames between I-frames, including the first I-frame. The group of pictures is shown in the shaded portion of Figure 2–10. The number of pictures in the group of pictures is called the *I-frame distance*; in this example, the I-frame distance is 6. The *P-frame distance* is the number of frames. In Figure 2–10, the P-frame distance is 3. The I-frame distance and the P-frame distance are important knobs. For television, which presents 30 frames per second, I-frame distance is on the order of 15, which means that an I-frame is sent every half-second. P-frame distance is on the order of 0–3.

Networking Implications

Frame architecture has implications for networking. I-frames anchor picture quality because P-and B-frames are derived from them. Therefore, it is important that I-frames be transmitted with high reliability, higher than P- or B-frames. Thus, when transmitting MPEG frames over ATM or frame relay, it is advisable that I-frames be given priority.

MPEG-2 creates streams of compressed video with timestamps. This timestamp is provided inside an MPEG-2 packet so that the timing and ordering of the I-, B-, and P-frames is correct. Note that because the MPEG-2 layer itself has a timestamp (Program Clock Reference, or PCR), it does not depend on the network layer to provide a timestamp for it. That is one reason why the ATM Forum decided that AAL5 would be used for carrying MPEG-2 streams rather than AAL1, which has a *synchronous residual timestamp (SRTS)*.

Encoder Operation

The job of the encoder is to compute compressed target frames from a perfect pixel-by-pixel image received by the camera. This image received by the camera is called the *original image*. Before any spatial compression is performed, the encoder caches the original image. As the encoder calculates the motion-compensated frame, it compares the work in process with the original frame. If the encoder decides that the original frame and the target frame are hopelessly out of sync (say, due to a scene change), then it can abort the process and immediately calculate a fresh I-frame. In addition to scene changes, motion compensation breaks can occur when background figures are covered or uncovered by foreground figures, or if extreme changes in luminosity exist.

MPEG-2 does not specify how an encoder works—it specifies only how the decoder works. The encoder's only explicit responsibility is to produce an MPEG-2–compliant bit stream; how it does so is up to the vendor.

The rocket science of encoders lies in determining as early as possible when the original and target frames are out of sync and then in taking proper corrective action within bandwidth requirements. Encoders can vary widely in determining when motion compensation is no longer required and in turning knobs other than recomputing I-frames.

Real-time encoding is made possible by the speed of modern processors. Real-time encoders require at least 1,600 MIPs processors.

MPEG-2 Systems Operation

Compressed video and audio streams are just the first step in defining a compression system. MPEG also defines an end-to-end software architecture for linking bit streams to programming and control information. This is defined in the MPEG-2 Systems specification, ISO/IEC 1383-1, "Generic Coding of Moving Pictures and Associated Audio: Systems," April 1995. The following sections overview the important components of MPEG-2 systems operation.

MPEG-2 Transport Stream Packet Format

The MPEG-2 Systems standard defined two data stream formats for performing systems processing in software: the *transport stream (TS)* (which can multiplex several programs and is optimized for use in networks where data loss is likely) and the *program stream* (which is optimized for multimedia applications), such as video storage on CD-ROMs or communications among video-editing systems. The TS is used for video over fiber, satellite, cable, ISDN, ATM, and other networks, as well as for storage on digital video tape and other devices. For purposes of RBB, attention has centered on the MPEG-2 TS as the format, so this section focuses on that aspect.

The basic difference between the transport stream and the program stream is that the TS uses fixed-length packets, and the program stream uses variable-length packets. The TS has a fixed length of 188 bytes. Table 2–5 displays field identifiers, the field length, and the function of the TS packet.

Table 2–5 *Transport Stream Packet Format*

Field	Length in Bits	Function
Synchronization	8	0×47
Transport error indicator	1	Determines that an unrecoverable bit error exists
Payload unit start indicator	1	Identifies first fragment of a Program Elementary Stream or Program Section
Transport priority	1	Gives priority to this packet over other packets with the same PID
Program Identifier (PID)	13	Identifies content, such as control table, or audio, video, and closed-captioned streams

Table 2–5   *Transport Stream Packet Format, Continued*

Field	Length in Bits	Function
Scrambling	2	Uses 00 means for no scrambling; the three other values are used and defined
Adaptation control	2	00 Reserved
		01 No adaptation field exists
		10 Adaptation field exists, but there is no payload
		11 Adaptation and payload exist
Adaptation field	256	See following fields
--Length	8	Determines length of adaptation field
--Program Clock Reference (PCR)	42	Provides timing information
-- Lots of Other Flags	Variable	

The adaptation field can be of length zero to 256 bits. If this field is greater than zero, bits are taken from the data portion of the packet, or the *payload*, to yield a total packet length of 188 bytes.

The fields with network implications are the PID and the transport priority. Certain PIDs and packets with high transport priority should be given preference by the network. All other fields are used primarily for decoding.

Program Identifiers (PIDs)

As noted in Table 2–5, 13 b of the header constitute the *Program Identifier (PID)*. The PID identifies the contents of a transport stream packet. Most TS packets carry digital audio, video, or closed-captioned text data for a TV program. The PID identifies the program for which the data will be used. In addition, four PIDs are used for control purposes, specifically 0, 1, n, and m. Table 2–6 summarizes these and other PIDs.

Table 2–6   *PIDs and Their Functions*

PID	Payload
0	Program Association Table (PAT), which points to the Program Map Table (PMT).
1	Conditional Access Table, which contains authentication information.
n	Program Map Table (PMT), pointed to by PAT, which binds PIDs to particular programs.
m	Network Information Table (NIT), an optional field that identifies physical transport, such as cable channel frequency, satellite transponder, modulation scheme, and interleaving depth. Its use is private and subject to the whims of the carrier.
0×0010 through 0×1FFE	User data, program content. These PIDs carry audio, video, and closed-caption data.
0×1FFF	Null packets, which are used for rate adaptation.

PIDs are of particular significance for channel selection. In the analog world, channel selection is performed when the receiver tunes to a single subband of a frequency multiplex broadband channel. In the digital world, channel selection is performed by joining a broadcast stream of digitally encoded packets. In the case of MPEG-2, tuning is PID selection.

A television program or a movie consists of at least three streams: a video stream, an audio stream, and a text stream, which contains closed-captioned text. Other streams might be associated with the program as well, such as foreign language audio. The streams are identified by the PID.

Each program therefore consists of at least three PIDs. Programs are mapped to PIDs through the *Program Map Table (PMT)*. When a consumer wants to view a channel, he consults an electronic program guide that displays what's available to watch and then makes a selection. In turn, a remote tuner selects the PIDs for decoding.

Figure 2–11 illustrates the interactions of the Program Association Table, the Program Map Table, and the transport stream.

Figure 2–11
PID usage.

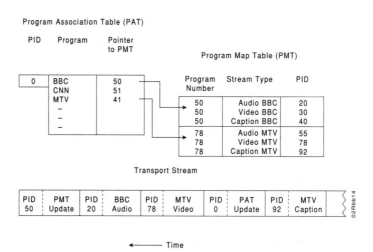

The figure shows a transport stream that consists of a multiplex of three TV channels: BBC, CNN, and MTV. The Program Association Table (PAT, PID=0) has an inventory of all channels available on the TS and a PID pointer to the Program Map Table (PMT) entry that identifies the PIDs of the data streams for each channel. In this case, transport packets with PID=50 will have information about the BBC, namely which other PIDs carry audio, video, and caption data for the BBC. Likewise, any TS packet with PID=41 carries PMT information about MTV. Note that PID=0 updates occur to refresh the PAT. As an example of standardization of control parameters, the Grand Alliance and the DVB Project have specifications on how frequently the PAT and PMTs are updated, in both cases at subsecond intervals.

Five transport stream packets are shown in Figure 2–11 and described in Table 2–7.

Table 2–7   *Transport Stream Packets*

Packet	PID Value	MPEG TS Payload
1	50	Contains update information for the BBC PMT. It, for example, might be necessary to change the BBC video PID.
2	20	Contains BBC audio. The payload contains MPEG-encoded audio for the BBC program. The set-top decoder that is tuned to the BBC would intercept this packet to render the audio feed.
3	78	Contains MTV video. The payload contains MPEG-encoded video for the MTV program. The set-top decoder that is tuned to MTV would intercept this packet and render it for viewing.
4	0	Contains a PAT update. Periodically, the PAT must be updated, so as to provide an accurate set of PIDs.
5	92	Provides MTV caption. The payload contains text information that is displayed for closed-captioning.

When a viewer wants to watch the BBC, he won't know that PIDs 20, 30, and 40 carry BBC packets. The decoder knows from the PAT and PMTs what PIDs to select.

Another use of PMT entries is to identify pure data streams, such as Internet service. The PAT, for example, can point to a PMT that identifies PointCast push-mode data, which can be distributed inside IP packets inside MPEG frames. Users who desire Point-Cast can select the corresponding PIDs just as they would select a TV program.

The example of Figure 2–11 also demonstrates that the stream of PIDs on a channel represents a multiplex of, in this case, two TV programs: MTV and BBC. Recall that a single 6 MHz channel can carry 27 Mbps. With MPEG-2, a program can use as little as 3–5 Mbps. Through the use of PMTs and PIDs, programs can be multiplexed on a single channel.

Another important notion is timing. Frames for a particular program must be displayed at the proper time, and the associated audio must be rendered concurrently. In Figure 2–11, for example, frames 3 and 5 (MTV video and captioning) must be displayed currently. Because the audio, video, and closed-captioned packets occur at various places in the multiplex, presentation timing must be coordinated. Frames 3 and 5 will have a *presentation timestamp (PTS)*, which indicates to the decoder when to display.

The PTS is an offset of the Program Clock Reference (PCR) in Table 2–5. Formally, an MPEG program is a set of packets that has a common PCR. In the multiplex of packets in Figure 2–11, for example, MTV has a different PCR from the BBC. That is, all MTV packets are relative to the MTV PCR, and all BBC packets are relative to the BBC PCR.

MPEG-2 Challenges

Although MPEG-2 compression has succeeded in producing high compression ratios (upward of 50:1), some caveats remain. These include picture quality, decoder costs, data services, low latency modes, and tighter integration.

Picture Quality

Both spatial and temporal compression produce signal loss. How well two different scenes will compress is not always obvious.

A cable operator testing digital video control parameters used an auto race and a fishing scene as viewing samples, for example. The auto race showed cars on a race track speeding by billboards. The fishing scene showed sunlight glistening off the water and leaves flapping in a breeze. Intuition suggests that the fast-motion racing scene would be more difficult to compress than the slow-moving fishing scene.

The fishing scene was very pretty in analog. In MPEG-2, the motion compensation broke down. MPEG-2 had problems with the leaves and the water because luminosity changed rapidly in a lot of places on the image. The result was a poor picture, even at relatively high bit rates. In the same test, auto racing looked quite good at a lower bit rate. MPEG-2 temporal compression accounts well for the movement of objects from frame to frame through the use of motion vectors, but there is no explicit temporal compression for quick and numerous luminosity changes from frame to frame.

The following list details scenarios that represent video compression problems for MPEG-2:

- Quick changes in luminosity at multiple places in the image, as demonstrated in the fishing scene. Flashbulbs also present problems.

- Circular motion, because motion compensation assumes that objects move in a straight line.

- Sharp, high-contrast edges for fonts and graphics.

- Multiple motions, where a single image splits into two or more, which confuses motion compensation.

- Alternating wavy lines, a variation on problems posed by circular motion.

Decoder Costs

B-frames impose a memory and processing requirement on the decoder. Most newer MPEG decoders with B-frames require more than 2 MB of memory. As higher-resolution video becomes popular, bandwidth restrictions might become even more important, which will tend to increase the use of B-frames and thus the cost of decoders.

Data Services

Data services such as web access and text annotation of broadcasts will be accommodated within MPEG streams. The adaptation of IP within MPEG is a work in progress and has not been fully addressed by the *Internet Engineering Task Force (IETF)*, which is the arbiter of IP enhancements.

Low Latency Modes

B-frames impose latency, which is not a big issue for broadcast, video on demand, or near video on demand. It is a problem for real-time video, such as video-conferencing. Low-latency forms of MPEG are needed to reduce or eliminate B-frames. MPEG has a low latency mode in which buffers are allowed to underflow for extended periods of time. When this occurs, the decoder might lose timing information until the underflow situation is corrected. Further work is in progress on low-latency MPEG.

Tighter Integration with Networking

A mechanism should exist by which the networking layer and the MPEG transport can help each other. A relatively straightforward case would be the case in which the network gives priority to I-frames and significant program identifiers. The network should be made aware of the existence of I-frames and give them preferential queuing. Preferential treatment also should be given to PAT and PMT updates, packets with PCRs, and frames explicitly marked for high priority.

SUMMARY

RBB investment and deployment will require familiarity with a wide range of technical subjects. New technology creates new businesses. Fiber optics, noise mitigation techniques, MPEG, and advanced data-networking techniques such as multicast have made RBB technically feasible. Table 2–8 summarizes the technical foundations of RBB covered in this chapter.

With the wide perspective on market drivers and the technology established in Chapters 1 and 2, the following chapters turn to a more in-depth consideration of the particular kinds of networks and their implications for RBB. In each case, technical, financial, regulatory, and market obstacles and facilitators of the networks will be considered. Cable networks are the first type to be covered in more detail, in Chapter 3, "Cable TV Networks."

Table 2–8   *Summary of Technical Foundations of RBB*

Technology	Factors Supporting Use in RBB	Factors Challenging Use in RBB
Metal wire transmission	Uses an entrenched, widely employed infrastructure	High-frequency transmission on metal wire is especially subject to impairments. Metal wire is prone to external impairments that cause noise.
Modulation schemes	Is a required aspect of communications; a number of robust, well-established, well-understood techniques exist.	Multiple modulation schemes exist; a relative lack of standardization that limits interoperability and raises costs.
Noise mitigation techniques	Combats the effects of ingress and burst noise, without incurring a round-trip delay. Is viewed as a required aspect of communications.	Creates delay.
Fiber-optic transmission	In comparison with metal wire transmission: increases bit rate; covers greater distance without amplification or regeneration of signal; lowers maintenance cost; requires less space; requires less weight; and is safer to handle. Is immune to eavesdropping.	Has unique impairments of its own, including: microfissures; difficulties in handling; inability to be bent like wire; and requirement of precise alignment of lasers required. Is immune to eavesdropping.

Table 2–8 *Summary of Technical Foundations of RBB, Continued*

Technology	Factors Supporting Use in RBB	Factors Challenging Use in RBB
Signaling	Is required aspect of communications. Is used with connectionless protocols, such as IP, minimizes signaling burdens, and promotes scalability. Packet networks in telephone systems enable faster signaling than previously possible.	Is time-consuming and imposes a processing burden on communications software. Can be the bottleneck that restricts bandwidth through networks.
IP multicast	Enables broadcast of data and video over wired networks. Is an important scaling technique for the Internet, because it reduces signaling and packet replication.	Uses complex software, requiring the cooperation of client system. Uses complex accounting and security. Scaling is difficult for multiple sources of multicasts. Is not widely deployed by Internet service providers.
MPEG-2 compression	Achieves substantial in storage and bandwidth requirements for audio/visual content; without it digital media would not be practical.	Does not work well for all video content; tradeoffs in visual quality are required. Makes networking more complex; requires integration with networking. Imposes delays.

REFERENCES

Books

Haskell, Barry, Atul Puri, and Arun Netravali. 1997. *Digital Video: An Introduction to MPEG-2.* Chapman Hall, ISBN 0-412-08411-2.

Varaiya, Pravin and Jean Walrand. 1996. *High Performance Communication Networks.* Morgan Kaufman Publishers, ISBN 1-55860-341-7.

Articles

[Cermak], Teare, and Stoddard Tweedy. "Consumer Acceptance of MPEG2 Video at 3.0 to 8.3 Mb/s," *Broadband Access Systems*, Wai Sum Lai, Sam Jewell, Curtis Stiller Jr., Indra Widjaja, Dennis Karvelas, Editors. *Proceedings*, SPIE 2917. 1996, pages 53–62.

"Generic Coding of Moving Pictures and Associated Audio: Systems," *Recommendation H.222.0, ISO/IEC 13838-1*, April 1995.

Balabanian, Vahe, Liam Casey, Nancy Greene, and Chris Adams. "An Introduction to Digital Storage Media-Command and Control." *IEEE Communications*, November 1996, pages 122–127.

Dutton, Harry. "High Speed Networking Technology." *IBM International Technical Support Centers*, GG24-3816-01, June 1993.

[Quayle], Alan, BT Labs "Telecommunications Operators Need and Requirements of an Interface between the Core and Access of a Broadband Network." *ATM Forum/96-1045*, August 1996.

Internet Resources

http://www.cisco.com
(Cisco's web page has many good White Papers. Use the search engine to find information on many topics.)

http://www.davic.org
(The home page of DAVIC.)

http://www-dept.cs.ucl.ac.uk/staff/jon/dummy/
(A characterization of distributed multimedia from Jon Crowcroft, University College, London.)

http://www.isi.edu/div7/rsvp/rsvp-home.html
(The RSVP home page written by University of Southern California Information Sciences Institute.)

http://law.house.gov/cfrhelp.htm
(The U.S. House of Representative Internet Law Library.)

http://www.light-wave.com
(Fiber news.)

http://www.mpeg.org
(The web home of MPEG.)

http://www.powerweb.de/mpeg/mpegfaq/
(MPEG FAQs.)

CHAPTER 3

Cable TV Networks

Present cable TV networks are shared, wired networks used primarily for one-way television transmission. In this case, the term *shared* refers to the fact that multiple households connect to a common piece of copper wire. This topology minimizes the number of route miles of wiring and is a natural topology for broadcast; but, cable networks require demultiplexing of traffic from different households to enforce security, provide bandwidth guarantees, and problem isolation. These problems are being addressed with new software techniques; cable networks are being upgraded for two-way transit and higher speed communications. In this revamped mode, plans exist to use cable networks for digital TV and high-speed Internet access.

Cable TV networks have the early lead over telephone companies and other service providers in offering broadband services in the home. Cable TV networks have speed, ubiquity, and experience in offering residential services, especially television. These advantages make it possible to offer digital television and high-speed Internet access to millions of consumers quickly over the existing network. Furthermore, cable TV is managed by executives with a history of entrepreneurial spirit who overcame great odds to create the businesses they have.

Their speed and multichannel potential raise the possibility that enhanced cable systems can be considered as platforms for innovative applications, such as digital storytelling. Time Warner Cable considers their new cable full-service networks as development platforms for such applications. They are particularly interested in the use of software objects to enhance the viewer's experience. Treating the cable network as an application platform fosters a new sense of creative development.

However, despite their apparent advantages, cable operators are faced with commercial and technical challenges as they try to keep their lead in the race of supplying broadband to the home. This chapter covers the technical and business obstacles that face facilitators of cable television's transition to residential broadband. It focuses on digital TV and data applications.

HISTORY OF CABLE

Community Antenna Television (CATV), commonly called cable TV, was invented to solve a dire consumer problem: poor TV reception. Rabbit ears and rooftop antennas don't suffice for people who live in valleys where nearby hills block good over-the-air reception, such as the upper Appalachian mountains. In fact, according to the Pennsylvania Cable TV association (http://www.pcta.com/histcabl.html), cable TV began in Pennsylvania. Others feel the CATV was born in Oregon. The following message was posted on the mailing list of the Society of Cable Television Engineers:

```
                    SITE
        OF THE FIRST COMMUNITY ANTENNA
           TELEVISION INSTALLATION
             IN THE UNITED STATES.
           COMPLETED, FEBRUARY 1949
                ASTORIA, OREGON

              CABLE TELEVISION
              WAS INVENTED AND
                DEVELOPED BY
              L. E. 'ED' PARSONS
              ON THANKSGIVING DAY
               1948. THE SYSTEM
             CARRIED THE FIRST TV
               TRANSMISSION BY
              KRSC-TV CHANNEL 5
            SEATTLE. THIS MARKED
               THE BEGINNING OF
                  CABLE TV.
```

Whether it began on the East Coast or West Coast, CATV was originally a small town phenomenon. The culture of rural independence and opportunism continues to differentiate this industry from its competitors.

Despite their best efforts, cable TV operators were a struggling lot from their beginning until the mid-1970s. The business was very capital intensive. The operator had to install thousands of miles of wires, erect buildings to house satellite and television transmission facilities, negotiate franchise agreements with municipalities, and conclude programming agreements before the first dollar was made. This was a risky investment proposition, especially because cable offered no distinctive programming at the time. It was funded primarily by subscription and local advertising, but there was no national advertising, and little transactional revenue from pay-per-view.

Furthermore, cable had touchy relationships with the cities in which it operated. To offer service, cable required franchises from municipalities in which it provided service. Until the 1980s, local authorities exerted strong control over cable operators. Cities charged large franchise fees, could require carriage of local broadcast, and regulated prices to their citizens. These were costs that over-the-air broadcasters did not have and put cable at a competitive disadvantage. But the difficulties in dealing with the cities were offset by the fact that the cable operator got an exclusive franchise to operate in the locality.

Even today, the subject of city-cable operator relationships continues to be sensitive. Franchise renewals are frequently challenged by consumers and cities that assert the cable operator has offered poor service or not fulfilled other provisions of their franchise agreement. Some franchise agreements included provision of public, educational, and government (PEG) access to channel capacity, increased bandwidth, and service level agreements.

So, undercapitalized and lacking compelling content that could differentiate cable from NBC, ABC, and CBS, cable did not catch on in a big way until the mid-1970s. In 1974, a technological innovation occurred that was to launch the cable industry out of its funk and into two-thirds of American homes today. That innovation was the commercial satellite; until then, satellites had been used exclusively for government purposes, mostly defense or space exploration.

With satellites, programs from multiple producers could be broadcast on a single transponder (aggregated) and rebroadcast to cable operators all over the country. This provided a means for producers to challenge the ironclad hold that the movie distributors had on the distribution of first run movies. Originally, satellite distribution was not directed at the broadcasters, but rather at the film industry.

The first company to aggregate and rebroadcast for cable distribution was Time Inc., which founded *Home Box Office (HBO)* in 1974. Viewers who wanted to watch HBO paid a monthly fee to watch programs they could not watch on over-the-air television. HBO caught on, and cable was on its way because it had content (movies) the broadcasters did not have, at prices that were competitive with or cheaper than movie theater tickets and video rentals. HBO is still the largest premium cable network in the United States with 19 million subscribers.

After the success of HBO, other programmers followed. A climate for more original programming was created and new channels carried only on cable, such as MTV, ESPN, and CNN, became part of popular culture both in the United States and abroad.

Cable had the advantage of greater channel capacity compared with over-the-air broadcasts. Some over-the-air bandwidth is set aside for use by police, fire, military, air traffic control, radio astronomers, and other public uses. The need to share the airwaves with public services, coupled with certain engineering issues, limits the number of over-the-air channels for a given metropolitan area to roughly 10–12. Instead of merely 10–12 stations, it was possible for cable systems to offer at least 30 channels over 300 MHz of cable bandwidth. Updated systems, using fiber-optic cables, have up to 750 MHz of bandwidth and offer over 100 channels.

See the web page for the *U.S. Department of Commerce Office of Spectrum Management (OSM)* at `http://www.ntia.doc.gov:80/ osmhome/osmhome.html` for details of spectrum allocation in the United States. The decision of who gets to use over-the-air spectrum is controlled by the Common Carrier Bureau of the Federal Communications Commission (FCC).

Because of superior reception, innovative programming, and increased channel capacity, cable operators succeeded in developing a large business since its inception in the 1950s. The following are some key cable statistics:

- Number of cable subscribers per region:
  - North America   73 million
  - Western Europe   43 million
  - Asia   19 million
  - Latin America   13 million
  - Eastern Europe   9 million
  - Worldwide cable subscribers   157 million
- United States infrastructure statistics:
  - U.S. cable system revenues $25 billion
  - Extent of U.S. cable   1.29 million miles of
    infrastructure   cable
  - Bandwidth available to   300–750 MHz
    each per subscriber
  - Number of TV channels   50–80
    available to U.S. subscribers
  - Subscribers per mile of   50
    cable in the United States

CABLE'S BUSINESS RATIONALE FOR RESIDENTIAL BROADBAND

Later sections in this chapter consider some of the business challenges and disincentives to cable's transition to residential broadband. The obstacles will be easier to understand after discussions regarding technical aspects of the upgraded system. Understanding the incentives, however, is pretty straightforward.

First, there is a defensive aspect to cable's motivation for transitioning to RBB. Residential broadband service is a necessary defensive measure against the threat of direct broadcast satellite and new telco services. In particular, DBS has signed over five million subscribers in three years, threatening cable's customer base for basic television and premium pay channel service.

Defensive measures aside, extending the capabilities of cable TV networks from the current broadcast video business to high-speed Internet access seems natural. Cable has a large capital base to use as leverage for incremental applications. In addition to being the incumbent, cable has specific technical and business advantages with which it can succeed in providing a wide range of high-speed home services. Among these advantages are:

- Ubiquity

- Speed

- Reduced signaling

- Tariffing

- Port conservation

- Vertical integration

Ubiquity

Cable service is available to nearly one hundred percent of residences in the United States and serves nearly 65 million homes. Worldwide there are 157 million subscribers. Widespread coverage helps attract advertising. In the United States, cable sells $6 billion in annual advertising for national and local spot ads.

Speed

Cable has more bandwidth than current telephone access networks and wireless networks. The speeds are so high that content providers who deliver over cable are considering changing their content specifically to exploit the speed of cable, for example, by adding more audio and visual content to web pages.

Reduced Signaling

Cable is always on; a call setup process, such as ringing a telephone, is not needed. The subscriber does not have to request that the cable operator transmit specific content, it simply arrives. Cable is basically in push mode, such as broadcast television. Reduced signaling is a prerequisite for widespread residential applications.

Tariffing

Because cable is always on, there is no time-dependent pricing. Pricing is subscription based (monthly access fee) or transactional (pay-per-view). This is predictable and user-friendly pricing.

Port Conservation

Because cable is a shared medium, the cable operator needs only a single physical port at the carrier premises to support hundreds or thousands of users. This is in contrast to telephone companies that must allocate a port for each subscriber. [Gillett] reported in 1995 that nearly a 30:1 cost advantage exists for cable compared with ISDN when measuring cost per bit of peak bandwidth per subscriber. Port conservation is an important cost advantage for cable in relation to telephone service.

Vertical Integration

Finally, the top cable operators in the United States are important content providers. TCI owns Liberty Media; Time Warner owns Warner Brothers, Turner Broadcasting, and a host of magazines (*Time*, *Life*, *People*, *Sports Illustrated*, and *Fortune*). When content providers own cable properties, they have guarantees of a high-speed outlet for their visual content, and plentiful content to occupy the many channels cable affords.

ARCHITECTURE

Figure 3–1 shows a schematic of a cable system that is not upgraded with fiber. Video programming for cable systems is obtained by satellite or microwave at places called *head ends*. Head ends have facilities to do the following:

- Receive programming (for example, from NBC, CBS, and cable networks such as MTV and ESPN)

- Convert each channel to the channel frequency desired, scramble channels as needed (for the premium channels)

- Combine all the frequencies onto a single, broadband analog channel (frequency division multiplexing)

- Broadcast the combined analog stream downstream to subscribers

The term *downstream* is used because the antenna has to be high enough to get clear reception, and signals are shipped downward to the viewers below. Downstream traffic is also referred to as *forward* traffic. There were 11,660 cable head ends in the United States as of December 1996.

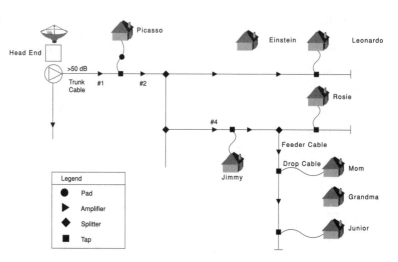

Figure 3–1
Cable system schematic.

The downstream video traffic emanates from the head end and is injected into a *trunk cable*, at a signal strength above 50 dB. Passive devices called *splitters* divide the traffic at branching points to provide geographical coverage. *Feeder cables* emanate from the trunk cables.

The head end and its connected coaxial cables and subscribers comprise a *cable system*. The operator of a cable system is a *cable operator*. Big companies are established that operate multiple systems. They are called *multiple system operators (MSOs)*. MSOs may own systems that are geographically contiguous or widely dispersed. Geographically contiguous systems offer a greater economic advantage than wide, dispersed systems. MSOs are aggregating their systems into large contiguous regions through mergers and acquisitions. The major MSOs in the United States are Tele-Communications Inc. (commonly known as TCI), Time Warner Cable (owned by Time Warner), Cox Cable, Comcast, Cablevision, and MediaOne (a unit of US West Media Group). Important MSOs in other countries are Canal Plus (France), Deutsche Telecom (Germany), and Rogers Cable (Canada).

Signal strength attenuates by the square root of frequency as it moves through the cables (both trunks and feeders). So, higher-frequency signals lose strength faster than lower-frequency signals. This is why programmers desire lower numbered channels. Signal strength is also attenuated by splitters and other passive devices.

To maintain signal strength, the signals are boosted with amplifiers roughly every kilometer (or even half a kilometer) if there are lots of subscribers. Because higher channels attenuate faster and the distances over which they must travel can be long, amplification for the higher frequencies might yield excessive power levels for low-frequency channels. So, there is a need to equalize power across the transmitted frequency band at the end points to reduce distortion.

Subscribers are connected to the feeders or trunks with the addition of a *drop cable* (300 feet of cable or less) and a connector between the drop cable and the feeder, called a *tap*.

Another approach to connecting users is to use long drop runs of low loss cables (more expensive than normal drop cables) to a central tap, and eliminate some of the distribution system. This improves reliability and reduces theft of service. It also substantially reduces the number of frequency limiting devices required, such as passives and taps. The drawback is a lot of potential signal loss due to the possibility of amplifiers being bypassed.

In the home there is the familiar *F-connector*, a cylindrical connector with a center pin sticking out, that is plugged in to a set-top box or cable-ready TV or VCR.

To cover a distant site, an amplifier might be turned up high, resulting in too much signal strength and distortion. So, nearby homes need a pad; a *pad* is a passive device that induces attenuation. For example, in Figure 3–1; to provide service to Leonardo, amplifier 1 is cranked up, thereby requiring Picasso to have a pad on his drop cable.

The result is a tree and branch topology with the head end as the root and the subscribers as the leaves. It is possible that the farthest subscriber could be 50–70 miles from the head end. Splitters and taps are necessary to distribute television signals, but have the side effect of weakening signal strength by introducing signal loss at each occurrence.

In Figure 3–1, eight homes are passed by cable and six are actual subscribers. A *home passed* is a residence that can be a subscriber with the addition of a drop cable. About 95 percent of all homes in the United States are passed. Einstein and Grandma count as homes passed even though they are not subscribers. The *take rate* is the ratio of homes that pay for the service to homes passed. In this example, the take rate is 6/8 or 75 percent. The take rate in the United States is 64.1 percent, as of December 1996, according to Nielsen Media Research.

HYBRID FIBER COAXIAL (HFC) SYSTEMS

Coaxial cable systems provided great improvements in television reception in many parts of the country. Even in areas where reception was good, CATV was popular because of increased programming options. Many new channels were premium or pay-per-view channels, thereby affording broadcasters new sources of revenue and content creators new outlets for their work.

Why Fiber?

Pure coaxial cable systems are insufficient in several ways for high-speed residential broadband service. First, channel capacity is still insufficient to compete with Direct Broadcast Satellite (DBS). Coax systems can provide roughly 40 channels. But DBS subscribers can receive twice as many channels, affording them more innovative programming options. DirecTV in particular has done well with enhanced sports packages. Cable networks require additional channel capacity to be competitive.

Second, coax systems lack robustness. If an amplifier malfunctions near the head end (for example, by losing power), all subscribers downstream from the bad amplifier lose service. Referring back to Figure 3–1; if amplifier 4 breaks, then Jimmy, Rosie, Mom, and Junior will lose service.

Third, signal quality is insufficient for large numbers of users. In Figure 3–1, Junior or Leonardo might be 40–50 miles from the head end. The number of amplifiers between the head end and the last subscriber is called the *cascade depth*. Junior is 6 amplifiers from the head end; that is, his cascade depth is 6. Cascade depths of 40–50 are possible in some systems in the United States. Amplifying a signal 40 times is like copying a tape 40 times. The last signal is never quite true to the original.

Finally, coaxial cable systems are *very* complicated to design and operate. Referring back to Figure 3–1, if many new subscribers living near Jimmy begin service, it is possible their taps would cause enough attenuation to require a new amplifier for serving Mom's neighborhood. But even if there were no changes in subscribership, there are dozens of amplifiers, splitters, pads, taps, cables of various impedances, thermal changes (which lengthen or shorten cables as ambient temperature changes), and pirates (doing their dastardly deeds between the head end and subscribers) spread over a 40–50 mile radius. Keeping power equalized for all subscribers is a difficult problem.

Ongoing operational network design is also complicated by rusting components, leakage due to perforation of insulation, temperature gradients, loosened fittings, and other annoying causes. These imperfections are difficult to find, and when found, typically require onsite maintenance. Each instance of onsite maintenance costs about $50–$100, which compares unfavorably to the $20–$40 month charges paid by subscribers. Software programs exist to assist technicians, but for the most part they use manual processes as well as a lot of intuition and experience.

To combat these problems, cable operators came up with the idea to use fiber-optic cable in place of coaxial cable trunks. The total system would have both fiber and coaxial cables, hence the term *hybrid fiber coax (HFC)* networks. With the invention of the linear light source and analog fiber systems, it became practical to introduce fiber into the trunks. The reason the fiber systems had to be analog was to maintain compatibility with the existing analog metallic plant. Also, the use of analog fiber was cheaper and more reliable than conversion to and from digital fiber.

By using fiber, the cascade depth can be reduced to a handful of amplifiers, say four or five. Reliability and equalization problems associated with deep cascades were reduced. The systems would offer more bandwidth and would be easier to design. Figure 3–2 shows how the system in Figure 3–1 would evolve with the introduction of fiber.

The diagram shows that trunk coax cables have been replaced by multiple fiber cables. If a fiber link breaks, fewer homes lose service. The fiber cables are terminated at media translators called *fiber nodes* that convert optics into electronics. Therefore, the cable system is segmented into smaller *clusters*, each of which is defined by those homes passed by a single fiber node. In this case,

Picasso, Einstein, and Leonardo are passed by a single fiber node and therefore are in the same cluster. It is the intent of MSOs to limit clusters to 500–2,000 homes passed.

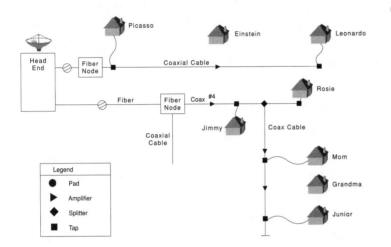

Figure 3–2

Hybrid fiber coax (HFC) schematic.

One to six coaxial cables are attached to the fiber node. In the forward direction, the fiber node splits the analog signal so that the same signal is sent on each coaxial cable. In the return path, traffic is received from the coaxial cables and multiplexed in the frequency domain onto the fiber. The definition and purposes of the return path are discussed in more detail in the section "Upstream Transmission," later in this chapter.

Between the fiber node and the subscriber, the assembly of coaxial cables, amplifiers, splitters, and taps is the same as the pure coaxial cable scenario. HFC systems reduce the number of amplifiers, thereby increasing available bandwidth and reducing maintenance.

A key innovation underpinning HFC deployment was development of amplitude modulated fiber-optic transmission. Until the invention of the linear light source in the 1980s, it was not possible to have *amplitude modulated (AM)* fiber; and without AM fiber, it was not possible to have analog transmission over fiber. Without analog transmission over fiber, cable would have had to perform analog-to-digital and digital-to-analog conversions, as discussed in Chapter 2, "Technical Foundations of Residential Broadband," to use fiber, a costly proposition. AM fiber was central to the evolution of HFC cable systems.

Mapping to the DAVIC Reference Model

Figure 3–3 shows how a cable HFC network maps to the DAVIC reference model introduced in Chapter 2. Content providers connect over the A9 interface to the Transport Network. The Transport Network connects to the Access Network over the A4 interface, which is normally ATM. One example of the access node of the DAVIC model is embodied in the *Cable Modem Terminal Server (CMTS)* for data and cable systems. The ONU is embodied in the fiber node. The A1 and A0 interfaces are embedded in the cable modem. The CMTS communicates with the cable modem to enforce the Media Access Control protocol and RF control functions, such as frequency hopping or automatic gain control.

Figure 3–3

Mapping to the DAVIC reference model.

Courtesy of Digital Audio Visual Council (http://www.davic.org)

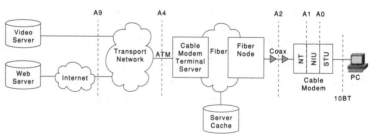

HFC Upgrade Costs

The cost of upgrading to fiber varies by geography, labor rates, and to some extent on what dollars are allocated to upgrade and what dollars are allocated to operational costs. Because metallic cables need periodic replacement anyway, it makes sense to replace metallic cables with fiber cables during normal maintenance. Over 90 percent of coaxial cable is reused to feed power to the fiber node.

The cost of upgrades includes the cost of optoelectronics (fiber, laser transmitters), changes to the cable plant (connectors, power supplies, passives, occasionally new cable), and labor. Such numbers are typically proprietary, but one study was presented in [Lyons].

The study involved upgrading to provide 750 MHz to clusters of 500 homes per fiber node. The resultant system provided 78 analog channels and 200 MHz of digital bandwidth. The costs were $16,732 per mile for aerial upgrade and $23,494 per mile for underground. These costs did not include return path or set-top costs. Given a take rate of 65 percent and a residential density of 80 homes per mile, this capital upgrade is $322 per subscriber for aerial upgrades and $452 for underground upgrades.

The study showed there is a point of diminishing returns for fiber upgrades. Extending fiber further into the network to 300 home clusters costs an additional 60 percent with no appreciable improvement in maintenance costs or reliability.

In addition to the capability to provide more channels, the benefit to the operator was greater reliability. Outages dropped by an estimated 75 percent, and there was a savings of $1,826 per mile of coaxial cable maintenance costs.

Cable operators around the world have seen the value in switching to fiber. They are upgrading to create clusters of 500–2,000 homes passed per fiber node as fast as their labor and finances permit. Given the nearly 1.3 million route miles of cable in the United States with only about 80,000 route miles of fiber in place, the upgrade to fiber is a multi-billion dollar effort. This expense is required if cable is to remain competitive within the home entertainment industry. For now, pure coax systems are still the dominant medium for CATV in the United States. As of December 1996, less than 20 percent of subscribers were connected through fiber nodes.

UPSTREAM TRANSMISSION

The introduction of fiber is the first step in cable TV networks' evolution to full-service residential broadband networks. The second major step is to create an upstream transmission capability in the HFC plant so that subscribers can send data as well as receive it. The terms *upstream*, *reverse*, and *return path* are used interchangeably to describe traffic and paths that go from the consumer to the head end.

Building on Existing Return Paths

Previously, many return path amplifiers were installed by cable operators to poll set-top units to collect *impulse pay-per-view (IPPV)* billing information. That information is retained in the set top and polled periodically by the MSO to obtain invoicing information. The return path is also a means for the operator to monitor the network. IPPV and monitoring are low bit rate uses of the return, so little return bandwidth is required. Today, with higher bit rate requirements for data, more return bandwidth and more aggressive modulation schemes are required than are necessary for simple invoicing data collection.

The vast majority of systems today assign upstream frequencies in the range from 5–42 MHz, below the frequencies used for broadcast television. Systems that split the upstream from downstream frequencies at frequencies below 54 MHz are referred to as *low split*.

Mid split systems separate the upstream from downstream frequencies above 100 MHz. Exactly where the return and forward paths split is decided by the operators. In a representative mid split, the return path is in the range of 5–108 MHz, and the forward path is 174–450 MHz, noted as 5–108/174–450. A *high split* system of 5–150/174–750 provides the greatest amount of return path.

Mid and high splits are useful for two-way communication, but infringe on frequencies used for legacy broadcast television. By starting the forward at 174 MHz, channels 2–6 are lost. Three broadcasters in any given market would lose their natural channel assignments. Consequently, low split systems are the focus for residential broadband rollout. Mid and high split systems are being considered for custom networks installed by operators specifically for industrial or governmental uses.

To provide reverse path or upstream support, amplifiers are needed on the coaxial cable in the upstream direction, and laser transmitters are needed in the fiber node to transmit back to the head end. Because return path frequencies are low, there is less signal loss in the return path than in the forward path, so fewer amplifiers are needed. The capital cost of return amplifiers is low compared with forward path lasers. Furthermore, low cost

Fabry-Perot lasers are typically used in the fiber nodes to illuminate the return path. Because of these relatively low cost components, the capital cost of upgrading to two-way is relatively low, about $10 per home passed.

In addition, labor costs must be considered. Estimates from some operators cost about $300 per mile to improve signal fidelity for return path. Improving signal fidelity means to clear away noise, tighten the cable plant against leakage, equalize frequency response, and perform other installation preparation. The cost per home passed is less than $10 which, like capital costs, is relatively low. However, the work is labor-intensive and requires skilled labor. Given the thousands of miles that need to be installed, labor availability is a restrictive factor in creating the return path.

Furthermore, after the return path is created, it requires significant incremental, periodic maintenance. Some estimates from operators indicate that there is an incremental requirement of one field technician per 200 miles of reverse plant. The availability of labor can be a significant problem, because telephone companies are sometimes competing for labor within the same pool.

The problems of impairments on a shared media are significant for both service initiation and ongoing service maintenance. Impairments are especially problematic for the return path. These will be discussed later in this chapter in the section "Technical Challenges to Data for Cable."

Telephone Return

Due to return path impairment problems, some cable operators are choosing to defer enabling the return path and opting for telephone return instead, at least for an interim period. By using telephone return, rather than cable return, two-way services for data and video on demand can be offered by the introduction of HFC only. Telephone return technology is well understood, and two-way service can be initiated by the MSO quickly.

Telephone return means that the consumer, or the subscriber modem, makes a telephone call to a terminal server when the consumer requires return path service. The terminal server may be located at the head end, a super head end, a telco facility, or a facility operated by an Internet service provider (ISP).

AT ISSUE

Viability of Telephone Return

Many observers contend that telephone return is not a viable long-term strategy for cable operators. It might not even be a profitable short-term one. Consider the following:

- The customer might pay for message units. Telephone return is advocated in geographical areas that do not justify return path buildout. But these areas, primarily rural areas, are also likely to be farthest from the dial point of presence, and therefore charged for message units.

- The consumer loses use of the telephone while it is being used for return path needs and might need an additional phone line to compensate.

- The consumer might get busy signals. In this case, there is no return path, so TCP acknowledgments can't reach the subscriber unit, which turns off all forward path traffic.

- Call setup is too long. Dialing takes a minute or so. This makes the TV viewing experience or Internet experience unsatisfying.

- For TCP traffic, the slow return path will limit the forward speed. The return path contains the acknowledgments that keep forward traffic moving. A 28.8 Kbps return path will limit the forward speed to 500 Kbps or so. This is a lot, but not the 27 Mb advertised for cable.

continues

- The cable operator must pay for the terminal server and access lines. If you count costs for rack space and management of multiple terminal servers, the cost differential between telephone return and enabling the cable return path narrows.

Because of the lack of satisfaction with telephone return and the relatively low cost of implementing two-way cable, two-way cable is expected to proliferate shortly. The investment firm of Goldman Sachs estimates that 40 million homes in the United States will be two-way–enabled by the year 1999, with Time Warner leading the way with 90 percent of their homes passed.

DIGITAL VIDEO SERVICES OVER TWO-WAY HFC

In 1994 when John Malone, CEO of TCI, talked about the 500 channel service, he was referring to MPEG-2–encoded digital TV over HFC. HFC systems would be built with 750 MHz of forward bandwidth, each offering 110 channels or more of 6 MHz. Each 6 MHz channel would be MPEG-2–encoded to yield 4–6 digital channels. Here's how.

Take 6 MHz and remove 500 KHz from each side for guard band, yielding 5 MHz. Use QAM 64 modulation on the 5 MHz. QAM 64 is selected by American and European standards bodies as the modulation scheme of choice for cable systems. QAM-64 codes 6 bits per hertz, yielding 30 Mb of raw throughput. Use Reed-Solomon forward error correction, which consumes 10 percent of raw throughput for error detection and correction, yielding 27 Mb. Twenty-seven Mb can comfortably support 4–6 channels of MPEG-2 programming.

It is believed that 750 MHz, shared among 500–2,000 households, would satisfy bandwidth requirements for legacy analog television, near video on demand, video on demand, Internet access, and the dozens of new networks and local or public service programming networks.

The next section covers how the digital video would operate and some of the issues involved.

Principles of Operation

Cable operators will be distributing analog and TV programs over HFC networks. Analog distribution is required throughout the period of transition from analog to digital over-the-air broadcasting. Digital channels will also be provided by broadcasters and other sources, such as independent producers who distribute over cable. Channel capacity will be divided between the analog and digital channels according to the availability of digital programming and management preferences of the MSOs. The analog channels will be carried from 54 MHz up to roughly 350–400 MHz. Digital channels will start at this cut off frequency and proceed to 550–750 MHz.

At the consumer site, an MPEG-2 set-top decoder transforms the digital transport stream so that an analog TV set can present the image. Analog programs will pass through the decoder directly to the analog TV.

The basic functions of the digital set-top box are:

- Diplex filters to separate upstream from downstream traffic

- Demodulation of downstream traffic from the HFC

- Modulation of upstream traffic onto the HFC

- An electronic program guide that displays programming options, includes virtual channels, and access to chat rooms linked to programs

- MPEG-2 PID selection

- MPEG-2 decoding

- Analog interfaces (RGB and audio) outlets to analog TV sets

- Descrambling of analog TV broadcasts

- Infrared controls (for channel selection) and VCR controls

- Transmission of return path signals

- De-encryption of downstream traffic

- Encryption of upstream traffic

- Authentication

- Separate, or out of band (OOB), control channel

- IP stack and IP address filtering

Optional components include:

- Interactive game interfaces, such as joysticks

- Data communications interfaces, such as Universal Serial Bus (USB), or Ethernet

- Web browser interfaces, to support interactive data services

- Java engines, which enhance web browsing and provide an application interface

- Smart cards for authentication

- Copy protection circuitry, to prevent unauthorized copying of digital content

Digital TV set tops will require use of the return path for user control purposes, such as channel selection, pay-per-view ordering and fault monitoring. There will also be applications such as WebTV that will have user interaction with a data-like application. Problems of return path are similar with data services over cable and will be discussed in the section "Technical Challenges to Data for Cable."

At the time of this writing, a rough estimate of the cost of elements of digital set tops were as follows:

Microprocessor, 4 MB of memory, flash memory	$100
Transmission Elements (tuner, equalizer, modulator, FEC)	$60
MPEG chips, graphics, and audio processor	$50
Chassis, power supply, final assembly, PCB, and test	$50
Analog and infrared interfaces	$10
Software licenses (OS, encryption)	$10
Total cost	$280

This total is the cost to the manufacturer; consumers would pay more, probably around $500. As time passes, the cost will decrease. A representative cable set-top box, to be implemented for Time Warner's Pegasus system, is described and illustrated in Chapter 8, "Evolving to RBB: Integrating and Optimizing Networks."

Challenges to Digital Video Service

As of 1997, the cable industry had yet to deliver on the vision of digital video service over two-way HFC. Meanwhile, the DBS industry has acquired over 5.1 million subscribers as of June

1997. In fact, the heaviest involvement of the cable industry in digital TV is the launch of their own satellite service, Primestar.

The challenges in the DBS industry are cost and standards. MSOs worry that decoders will be viewed as prohibitively expensive by many consumers; they would like to see set tops priced under $500 (despite the fact that DBS originally priced their set tops at nearly $1,000, including installation). Standards for modulation schemes, forward error correction, and authentication have only recently been agreed to.

In addition to cost and standards, availability of components has been something of a roadblock. Only recently have low-cost, QAM-64 ASICs been made available.

Because of these and other difficulties in offering digital video and derivative services, such as near video on demand, some MSOs have been pursuing Internet access with greater vigor.

DATA SERVICES OVER HFC SYSTEMS

The second major market for two-way cable is data, such as Internet access. The idea of moving data over cable TV networks is an old one. Cable modems were merchandised as early as 1987 by companies such as Fairchild Electronics. The Fairchild M505 Broadband Modem was QPSK and QAM-modulated, yielding up to 10 Mb of full-duplex service. Many of its characteristics are surprisingly similar to the cable modems being developed today. The idea then was to use coax networks primarily for intranet use. So, the 1980's generation of modems lacked the scaling properties necessary for public carrier use.

With the growth of the Internet, cable operators have begun thinking of their HFC networks as a logical vehicle for improved data service. Internet access is an inherently asymmetric service, meaning more data comes from the head end than to the head end, a characteristic that matches the bandwidth capabilities of low split cable networks. Furthermore, cable networks have the ubiquity to quickly offer service to large segments of the population. Finally, the convergence of Internet access with digital TV made it necessary for cable operators to offer data service and to prevent cannibalization of their video service.

For these reasons, cable has significant leverage and motivation to deploy residential data service. Various research groups are predicting big things for broadband data service. The following are variables affecting the sales of cable modems:

- How many systems are upgraded to fiber.

- How many systems are two-way–enabled.

- How well operators can recruit and train field engineering talent.

- How well MSOs can raise capital to recruit and train field engineering talent.

- How well web servers can deliver significantly faster service to showcase the speed of cable networks when compared with their telco counterparts.

- How popular competitive technologies will become, including ISDN and 56 Kbps modems.

- Pricing of high-speed services by Internet service providers.

- Pain threshold: How patient will cable operators be if their services do meet with initial consumer resistance?

Given these variables, estimates of data and cable penetration vary widely. Table 3–1 is a list of forecasts for cable modem subscribers by the year 2000 in the United States, estimated in 1996 and 1997 by various industry analysts. There are forecasts that are more optimistic and more pessimistic; this is a middle-of-the-road group.

Table 3–1   *Estimated U.S. Cable Modem Subscribers by the Year 2000*

Industry Analyst	Estimated U.S. Cable Modem Subscribers
Paul Kagan	3.8 million
Probe Research	3.4 million
Ovum	3.2 million
Simba Delphi	1.9 million
Kinetic Strategies	1.6 million

Service revenues are in the range of $300–$500 per subscriber per year. This yields annual revenues of $500 million–$2 billion per year.

Though substantial, this is far less than U.S. cable's current revenue of $25 billion per year.

Principles of Operation

The reference architecture discussed here is derived from the *Multimedia Cable Network System Partners Ltd. (MCNS)* partnership, a group of North American cable operators who is writing a purchasing specification for cable modem systems.

The components for a cable modem are shown in Figure 3–4. They are similar to the components of a digital set top, but without the MPEG elements. Because it does not include MPEG elements, the cable modem is cheaper than a digital set-top box. Approximate cost as of this writing is:

Microprocessor, 4 MB of memory, flash	$100
Transmission Elements (tuner, equalizer, modulator, FEC)	$60
MAC ASIC	$20
Ethernet/ATM 25	$10
Chassis, power supply, final assembly, PCB, and test	$50
Software licenses (OS, encryption, networking)	$20
Cost of Goods Sold	$260

Again, this is the cost to the manufacturer. Consumers can expect to pay $450–$500. As with digital set tops, the costs will decrease over time.

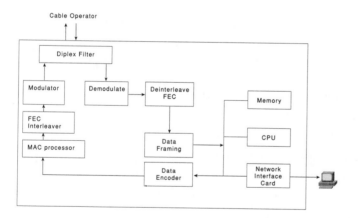

Figure 3–4

Cable modem schematic.

The operations of data services over HFC will be considered here in two basic steps:

- **Startup.** What happens when a new modem is attached to the network?

- **Data exchange.** How a modem sends and receives data.

Startup

When a new cable modem is installed on a cable network, there must be a very user-friendly installation process. Basically, everything needs to happen automatically. The modem needs to know which frequencies to listen to and on which frequency to transmit for downstream and upstream data, respectively. It needs networking information as well, such as an IP address and packet filters.

The startup process begins with the cable modem scanning all downstream frequencies it can for standard control packets it recognizes. These packets contain messages that are broadcast downstream, expressly for newly attached modems, from the *Cable Modem Terminal Server (CMTS)*. In the broadcast, there will be a command to transmit on a specified frequency.

Control packets, such as system initialization packets, are not encrypted. It would be impossible for an uninitialized modem to sign on to the system without prior knowledge of the encryption keys.

After the modem knows which upstream frequencies it will use, it begins *ranging*. Ranging is the process by which a cable modem learns its distance from the head end. This is required for timing information; that is, for synchronization purposes. All MAC protocols have a system-wide view of time, so that each modem knows when to transmit. Ranging is accomplished by the cable modem's sending a short message to the head end and measuring the time interval until it receives the message.

Ranging is a continual process because of thermal changes that cause the cable to expand and contract during the day. The thermal problem can become particularly acute. A *suckout* occurs when the center conductor contracts, due to thermal contraction, such that it disconnects from the center seizure of the coax connector.

System Resets

The CMTS requires a mechanism to constrain the number of cable modems that access the system during initial ranging to prevent upstream receiver saturation. Saturation can occur after a system-wide outage, such as a power outage or the failure of a fiber node. Unless there is an orderly scheduling process, ranging can take minutes after a system-wide reset.

IP Address Acquisition

After the ranging process, the cable modem is ready to obtain an IP address and other network parameters. It would be customer-unfriendly and create security problems if the customer were to configure his own IP address.

The modem obtains an address by using the *Dynamic Host Configuration Protocol (DHCP)*. DHCP is the standard Internet protocol for dynamic assignment of IP addresses. It is described in RFC 1531 (DHCP Protocol), RFC 1533 (DHCP Options), and RFC 1534 (DHCP/BOOTP Interoperation; BOOTP is the predecessor to DHCP). When a subscriber requires an address, the cable modem launches a particular type of broadcast packet, called a DHCP request, onto the return path. A router at the head end receives the DHCP request and forwards it to a DHCP address server, which is known to the head-end router by configuration. The server returns an IP address to the router, which caches it and relays the information to the subscriber. In addition to storing the subscriber's assigned IP address, the router will try to identify and store the MAC address and possibly modem serial number of the subscriber. Thus, the router keeps a database of all necessary address bindings for a particular user.

Because a DHCP handshake occurs prior to data transmission, it makes sense to enhance DHCP to exchange service information. The DHCP address server, for example, can be linked with a subscriber management database. Subscriber management can provide bandwidth guarantee parameters, (semantics to be determined), packet filters (semantics to be determined), and graphical user interfaces (customized per user). Work is proceeding in the IETF in the *IP over Cable Data Networks (IPCDN)* ad hoc group to define a version of DHCP customized for cable systems.

IP Subnetting

All data subscribers downstream of one CMTS RF port will be part of one IP subnet. Other subnets may be located behind downstream routers, but those routers' HFC interface addresses will be on the main subnet. Hosts on an Ethernet subnet expect to ARP for other hosts on the same subnet. However, the physical media does not allow one subscriber to communicate directly with another—subscribers can only send to the head end. Therefore, the head-end router must make a proxy ARP reply on behalf of the target host. Because these hosts are on the same subnet, normal proxy ARP rules would prevent a reply in this situation. Some special code will be required to modify this rule for HFC subnets.

Data Exchange

After ranging, the determination of upstream and downstream frequencies, and the acquisition of an IP address, the cable modem is ready to exchange data. Figure 3–5 shows a schematic of the path and process the data follows.

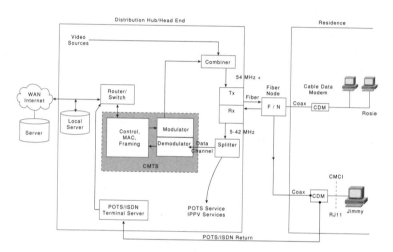

Figure 3–5

Data and cable schematic.

Starting from the home (right side of Figure 3–5), the PC generates upstream traffic and sends it over a network interface card (NIC) to a cable modem. Return path is activated so that data moves from the cable modem to the fiber node in the neighborhood.

The issue at the fiber node is that there might be multiple coax links feeding it. If there is excessive power generated electrically from the coax, as happens with packet collisions or the simple statistical alignment of waveforms from multiple sources, their summed power might overdrive the laser transmitters, causing laser clipping.

Eventually, traffic is received at the head end. Some of the upstream traffic is split off for other purposes, such as pay-per-view, telephony, and monitoring. The remaining data traffic is sent to a demodulator, which then sends bits to a terminal server, indicated as the Cable Modem Terminal Server (CMTS) in Figure 3–5. Traffic is then sent to a local router or switch for transmission to a local server, the Internet, or another cable subscriber. If the cable plant is not capable of two-way service, telephone return can be used. Jimmy's cable modem in Figure 3–5, for example, has an RJ11 jack for this purpose.

In the forward path, data is received at the router/switch and sent to the CMTS. Traffic is modulated, sent to a tuner so that the data can be on the frequency expected by the subscriber modem, frequency multiplexed with standard video traffic (in the combiner), and transmitted to the distribution plant.

When it wants to send data upstream, the cable modem complies with a Media Access Control (MAC) protocol that arbitrates contention on the upstream path among multiple cable modems. There is more on MACs later in this chapter.

How much of the 5–42 MHz to use for the return path is not standardized. Motorola uses as little as 600 KHz, Zenith and LanCity (a division of Bay Networks) use 6 MHz. The advantage of using a narrow return path is that there is a relatively flat frequency response. Also, with narrow return paths, if excessive ingress noise is a problem, the return path can be relocated elsewhere in the 5–42 MHz range. The downside is that there is a lower bit rate, using comparable modulation. MCNS and other forums are working to standardize return path channelization.

On the return path, most vendors choose to use QPSK modulation because of the noise resilience of that scheme. If the cable plant is relatively benign, QAM 16 can be used. QAM 16 doubles the data carrying capacity over QPSK.

Cable modem vendors agreed to use an entire 6 MHz channel for the forward path. This enables the MSO to use the same RF multiplexing equipment for data as they use for video. Six MHz of bandwidth on the forward allows 27 Mb of data per channel, using QAM-64 modulation and RS FEC. The 27 Mb are shared among all subscribers in a system. If there are 27 subscribers online at any instant, each can expect to receive 1 Mb of service, far exceeding the capability of telephone modem and ISDN service.

Table 3–2 tabulates the characteristics of the cable plant for data.

Table 3–2   *Return Path and Forward Path Characteristics*

	Return Path	Forward Path
Also Known As	Reverse path, upstream	Downstream
Channelization	Varies by vendor	6 MHz
Modulation Scheme	QPSK, QAM 16	QAM 64
Access	Via MAC protocol	Broadcast
Shared bandwidth	768 Kbps–10 Mb, varies by vendor	27 Mb

Owing to recent interest in high-speed Internet access, many companies—some think too many—are developing cable modems. Note that in 1997 IBM, Hewlett-Packard, and Intel terminated their cable modem production plans. Cable modem vendors are addressing difficult technical problems.

Technical Challenges to Data for Cable

Widespread industry development of cable for data depends on resolving several technical challenges:

- Return path noise problems

- Problem isolation

- Media Access Control protocol (MAC)

- Last mile framing

- Scaling techniques

- Security

Return Path Problems

The return path is enabled in the frequency range of 5–42 MHz. This is low frequency that has good attenuation properties. On the other hand, because there is heavy use of those frequencies by other services, ingress noise is a problem.

More importantly, the shared nature of the return path creates a funneling effect. Impairments are received by coax wires (which act as giant antennas) in the home and sent upstream. These impairments are funneled together as the signals move upstream owing to cable's tree and branch topology. This means noise gets increasingly louder as you go upstream, rendering the signal potentially indecipherable. For a summary of issues of return path problems, see [Prodan].

Sources of Ingress

Ingress noise is picked up by the cable from outside sources, through leakage and bridged taps. Here's a partial list of problems that result in ingress noise:

- HAM radio and citizen's band (CB) radio are allocated frequency by the FCC within the 5–42 MHz passband.

- Poorly insulated equipment in the home, such as microwave ovens, televisions, radios, and electric motors (vacuum cleaners, and so on). If these are operating near a loose F-connector, the noise will be picked up on the cable and sent upstream.

- Emissions from outside the home, such as lightning, neon lights, vehicle ignitions, and power line interference.

- The output of digital signals from the home terminal. Sometimes the source of noise could be the cable modem or the digital set top itself.

- Tampering and malice.

- Loose F-connectors in the home.

- Loss of integrity of the reverse optics, trunk cable imperfections due to corrosion. The design life of taps is 15 years, but in reality, street taps don't last longer than five. Acidic content in soil can break down the insulation between the cables and taps.

Despite the miles of cables owned by the operator, noise characterization studies have determined that most ingress noise on return path cable plants comes from inside consumer's homes, which is out of the control of the cable operator. The problem is, which home? Cable is a shared medium, which makes identification of the source of noise difficult.

Common Path Distortion

Another return path problem, frequently mistaken as ingress noise is *common path distortion (CPD)*. CPD is the interference of return path signaling caused by the forward path. One specific source of CPD is corrosion on the forward line. The corroded metal acts as a mixer diode, creating radiation on the reverse path from the forward carriers. Another possible cause is insufficient isolation between the downstream amplifier module and the reverse amplifier module in the same housing. It is possible that the isolation in the diplex filter is insufficient.

Power Management

Besides being subject to ingress noise and common path distortion, the return path presents a power management problem.

Power management of a two-way cable plant is a new technical problem. Previously, with one-way plants, there was no need for remote control of the output transmission power of subscriber units. In a two-way plant, careful control of transmit power on the return path is required. If one subscriber unit transmits with too much power, other subscriber units cannot be heard at the head end.

Mitigation Techniques for Return Path Noise Problems

A number of techniques are used to combat return path noise problems; these include filtering, automatic gain control, frequency hopping, and spread spectrum.

Filtering

The effect of ingress noise on one drop cable is exacerbated because the return path is a shared medium. If there is a strong source of interference from one household, it will pollute the spectrum for the entire neighborhood. Thus, a single household can cripple service for an entire neighborhood, even if the offending household subscribes to only the one-way service and not the two-way service. Mechanisms must be employed to identify and neutralize this problem.

One way to protect the upstream is to install a complete blocking filter near every tap in the neighborhood. The filter is inexpensive, and the labor is inexpensive because it is installed unconditionally for every cable subscriber. The filter completely disables all return path transmissions from every home. When a customer subscribes to a two-way service, the filter is removed and a *low pass filter (LPF)* is installed on the customer's cable drop, which requires the return path. The low pass filter permits low frequencies to pass, but blocks higher frequencies.

If, for example, a customer subscribes to cable data service, the LPF would be put on the cable drop in the home for the computer that allows data from the computer to pass to the head end. There will be a blocking filter for the cable drop to the TV to prevent the TV from polluting the upstream path for the neighborhood. If the consumer decides to move the PC and TV, then the LPF and the blocking filters might need to be reinstalled.

It is still possible that one of the two-way subscribers will induce ingress noise, but this technique ensures that the offending household's noise will be limited to those who actually subscribe.

Frequency Hopping

Using *time division multiple access systems (TDMA)*, there is the possibility that narrowband interference will become so severe for a long duration that a particular return path frequency will become unusable. Reasons for this include amateur radio, AM radio, or other stationary sources. In this event, the subscribers' modems must be directed to move to another transmit frequency on the return path en masse. This is accomplished by the CMTS sending a control message to the cable modem that identifies the new transmit frequency.

Hopping is made difficult when return path channelization is in a relatively broad frequency range. Cable modems from LanCity and Zenith, for example, use 6 MHz of return path. When hopping is required, they will need to find another contiguous 6 MHz chunk of bandwidth. Sometimes it might not be available.

Other modems use narrower bands. Motorola uses 600 KHz of bandwidth for the return. This makes it easier to find bandwith on which to hop. But, this severely restricts the bit rate available to the subscriber on the return. There is a fine balance between bit rate and reliability when sizing the return path.

Hopping solutions work well for narrowband interference but will not be effective to combat wideband impulse noise. For these problems, forward error correction or spread-spectrum techniques are required.

Spread Spectrum

Because the return path resembles a hostile jamming environment, spread spectrum is being considered as an alternative to frequency hopping. By using spread spectrum and a *code division multiple access (CDMA)* technique, proponents assert that spread-spectrum cable modems can operate at significantly degraded signal to noise ratios.

Another advantage of spread spectrum is that it tends to reduce the incidence of laser clipping, because less coincidental peaking of energy and phase by multiple transmitters occurs.

Spread spectrum is not appropriate for forward traffic because the noise environment is less hostile and it is viewed as a complex and expensive technology. Implementations of spread spectrum have been proposed by Unisys, which is no longer active in commercial cable modem development, and the Terayon Corporation (http://www.terayon.com).

Automatic Level Setting

Automatic level setting (ALS) mitigates the power management problem posed by the cable return path. It is the process by which the transmit power of the cable modem is regulated by the CMTS. ALS differs from *automatic gain control (AGC)* in that AGC doesn't typically employ remote sensing. When a particular cable modem is transmitting at inappropriate power levels, the CMTS senses this and sends a message on the forward path to the cable modem to either increase or decrease power as needed. In extreme cases, it will be necessary for the network operator to remotely disable a screaming cable modem.

Problem Isolation

Automatic gain control is a specific case of the difficulty of problem isolation on a wide area over a shared medium. If Jimmy's modem has a malfunctioning transmitter, it requires special processing for the head end to determine immediately if Jimmy, Rosie, or Junior is the actual culprit because all their traffic arrives at the head end through the same port.

Media Access Control (MAC)

Because cable systems are shared media, there needs to be an arbitration mechanism to decide who gets to use the medium at any instant. This mechanism is the Media Access Control (MAC) protocol. Because the downstream path is broadcast, there is no problem deciding who uses it. The MAC is needed only on the return path.

If Junior and Rosie decide to transmit at nearly the same time, their bits will collide rendering both transmissions indecipherable when they reach the head end. There needs to be a way for them to decide which one goes first and for how long before relinquishing the upstream to the other, or to someone else.

The objectives of any MAC are to do the following:

- Support the maximum number of possible users per cluster

- Minimize latency on data transfer

- Provide fairness, so that all users get some access

- Achieve maximum bandwidth utilization of the return path

- Evolve to future uses

- Support multiple classes of service

- Optimize costs

Optimizing costs is achieved by moving as much complexity as possible from the subscriber modem to the head end, to keep the subscriber unit as simple and cheap as possible. It is against these criteria that MACs are evaluated.

Structure of Media Access Control for HFC

Well-known LAN schemes such as Ethernet and Token Ring are not being considered in the cable environment, primarily due to their distance limitations. If Rosie was 10 miles from the head end, a round trip would take 120 microseconds. At 10 Mbps, that would mean 1,200 bits would be in flight before a collision can be detected. A short message could be sent by another user in the interim. Instead, in Ethernet or Token Ring set-ups, Rosie and everyone else would have to wait for the packet to pass. Polling techniques, such as Token Ring, consume excessive amounts of time for the polling sequence. Polling has never been considered by data and cable vendors.

Instead of using standard LAN techniques, vendors are converging to credit-based schemes. With credit schemes, the CMTS issues credits to the subscriber modems, granting them access to the upstream for a specified period of time. The time interval at which the cable modem is allowed to transmit is limited by their grants. After grants are exhausted, the cable modem must stop transmission. Variations exist as to the amount of data transmitted per grant (ATM cell or not), the number of credits issued by the CMTS per grant request, and backoff procedures in the event of collision of grant requests. These variations mean various modems from different vendors are not interoperable. Standards bodies, however, have come to a consensus on the direction that MAC development should be taking.

Performance of Grant Requests

Credit mechanisms must ensure that the CMTS can provide grants fast enough to keep bandwidth utilization high. If grants from the head end come to the subscribers too slowly, the subscriber modems stop transmitting. The result is that the return path is underutilized. Because the return path contains acknowledgments that permit forward path traffic, the forward path might become underutilized as well.

Normally, only one request may be outstanding from a single subscriber at a time. With a 4 ms interleaving delay on the forward path, and a 1.6 ms round trip propagation delay and processing time, it can take over 6 ms from the time a request is made by the subscriber to the time the grant is received. Efficient MAC algorithms are required to ensure that grant requests are expedited.

Collision Resolution Algorithms

Collision Resolution Algorithms (CRA) define procedures for subscribers to resubmit grant requests after a collision has occurred. Keep in mind that collisions only occur on grant requests. Data packets do not collide because they are transmitted by the subscribers under strict orders from the head end, which preclude collisions.

When subscribers collide, they are unaware of the event unless informed by the head end. The head end detects the collision and informs the cable modem that the collision occurred. One simple method of CRA is based on a truncated binary exponential backoff algorithm (basically Ethernet).

When informed of a collision, the cable modem randomly selects a number within a range called the backoff start window. The selected random number is a time interval for which the cable modem must wait before attempting to resubmit the grant request. If the backoff start window is sufficiently large, then the contending modems will be less likely to pick the same number and collide again.

If they collide again, the backoff window is increased by a factor of two (hence the term binary) and the contending modems repeat the process. There is a maximum to the backoff start window. If there are repeated collisions at the maximum, the contending modems will continue to retry until a maximum number of attempts as specified by carrier network management is reached, after which the sending data is discarded.

If, during this retry process, other subscribers submit grant requests, things get complicated. One approach to simplify this process is to enforce a policy whereby only the original colliding parties are allowed to resubmit requests.

Applying CRAs to actual systems introduces a lot of variables, which makes them difficult to test. For example, performance varies depending on admission control strategies, the number of active users, piggybacking, size of the request, and end-to-end delays in the system. These factors preclude MAC protocols from being analyzed. Real implementation with a large number of users has yet to be performed for most MACs, except by simulation. See [Golmie] for an interesting study comparing ternary tree versus adaptive p-persistence algorithms.

MAC design on shared media with large propagation delays remains an interesting research topic and the subject of intense competition.

Last Mile Framing

Last mile framing refers to the data encapsulation and transmission protocols used between the consumer premises and the head end. The candidates are ATM, MPEG, and IP. Each framing technique has its adherents.

Some argue that ATM should be used because it is most amenable to bandwidth guarantees, which is necessary on a shared medium. ATM cells are short, which means there is less likelihood of a collision on grant requests. ATM is proposed as mandatory by the IEEE 802.14 committee.

Others argue that MPEG transport streams should be used. MPEG-2 is already the framing technique used for digital video, and extending it for data is relatively straightforward. Simply encapsulate an IP packet inside one or more MPEG TS frames, and identify a PID for data. The set top can be used to pick off the data PIDs and send them to an Ethernet port on the cable modem.

Conversely, an IP packet can carry an MPEG TS frame inside. This would require the set top to be aware of IP packets, remove the IP headers, and decode the MPEG. Framing will be discussed in Chapter 8 as well.

[Adams] presents an overview of Time Warner Cable's full service network in Orlando, Florida, including how last mile framing was addressed. Orlando was originally a system that used ATM to the set top to provide broadcast and interactive services over a common HFC network. This meant that the set top performed reassembly of MPEG-2 from ATM AAL5 cells and performed physical layer functions such as FEC, encryption, and modulation. ATM was a logical last mile framing technique because it could handle data, video, and telephony traffic.

However, after initial deployment of thousands of units, it was decided that it was preferable to terminate the ATM at the head end and use MPEG to the consumer.

Scaling Techniques

Because cable is a shared medium, care must be taken to avoid congestion as the number of users grows and the usage of each user grows. In both cases, there needs to be a plan for how to manage scaling for video and ATM for data services and control functions.

Multiple Return Paths per Forward Channel

In the event that return path traffic grows, collisions might increase without a corresponding increase in forward traffic. In this event, it might be necessary to add an additional return path to complement the forward channel. Now there are two return paths, each with its own port.

In Figure 3–2, for example, Jimmy and Rosie share a common fiber node on the return path. Because the return is used heavily, the cable operator could assign Jimmy and Rosie to different frequencies for the return path, even though they use the same forward channel. Control messages from the CMTS must instruct the respective modems to use their assigned frequencies. The CMTS is responsible for load balancing across the return paths.

Having multiple return paths potentially complicates the MAC design process. Data received on two different ports at the same time is not a collision, but the MAC will still consider it a MAC collision unless proper care is taken.

Using Additional Forward Channels

A single forward channel provides a shared 27 Mb service to a neighborhood. When take rates and individual utilization permit, another forward channel can be allocated to data service. This means one less channel for analog data or multiple digital TV channels. The tradeoff of bandwidth between data and broadcast use is strictly an economic tradeoff. Again, the CMTS must perform load balancing, this time on the forward path.

Smaller Clusters

A final scaling technique is to extend fiber farther into the neighborhood. Initial rollouts of two-way services may be limited to clusters of 2,000 homes passed. In response to congestion, it might be necessary to reduce clusters to 1,000 or even 500 homes passed. Unlike using multiple return channels or additional forward channels, this step requires capital expenditures. So for those MSOs that don't commit to small clusters early, this is likely to be the last resort.

Security

Because HFC is a shared medium, everyone in a cluster can, in principle, see everyone else's data or, in the case of premium programming, receive programming without paying for it. Showtime Entertainment estimates the rate of theft at $100 million annually. *Communications Engineering Design* magazine reports 53 percent of systems surveyed filed charges for service theft during 1995 and that systems operators would be willing to spend up to $40 per set top for improved security for video service [CED January 1996]. The cable operator is required to offer an encryption mechanism to prevent unauthorized reception or modification of subscriber data. The main features of HFC security are the following:

- Having a proper authentication mechanism to ensure that the cable modem is entitled to service and has proper filtering and accounting profiles.

- DES encryption using RSA or Diffie-Hellman keys stored in the modem. Keys may be changed as frequently or infrequently as desired.

- All cable modems have a unique signature (or MAC address) that can be encrypted and used for authorization of the modem, or passed via telephone when the subscriber requests service.

- Decryption delays preclude encryption of MAC messages to assure that real-time MAC messages can be handled as quickly as possible. This means control messages are decrypted, which poses risks on shared media.

These measures are in addition to any application level end-to-end security. Security is the subject of discussions in various standards organizations and remains an open issue.

Security vulnerabilities are specific to user environments. For example, Windows 95 users on cable data systems have been advised to disable file sharing in the Networking menu of Windows 95's Control Panel file. Otherwise, it is possible for others on the cable system to peer into the disk drives of the exposed user. However, turning off file sharing creates problems for users who have local area networks. Additional software is needed to provide security on cable systems while enabling file and print sharing.

It should be observed that security measures for data are fundamentally different from security measures for digital video or analog scrambling. Security for data relies on unique descrambling or de-encryption per user session. Broadcast security cannot be done this way because all viewers are receiving a common program. Therefore common descrambling or de-encryption is required. This means that data security is structurally different from broadcast security and dual security mechanisms are required.

STANDARDS ACTIVITIES

A principal objective of the cable industry is to encourage retail distribution of cable modems and digital set tops. This removes inventory from MSO books and scales distribution and support. So far, no single standard for either of these components exists. Standardization is needed to reduce time to market, create a more competitive environment within the vendor community, reduce cost, and improve ease of use.

Multimedia Cable Network System Partners Ltd. (MCNS), IEEE 802.14, DAVIC, Digital Video Broadcasters Project (DVB), Video Electronics Standards Association (VESA), Society of

Motion Picture and Television Engineers (SMPTE), IETF, International Telecommunications Union (ITU), and the Society of Cable Television Engineers (SCTE) are working to standardize various aspects of digital and two-way cable, including physical layer (PHY), MAC, security, and service management for digital video and data. Their joint work is yielding a rich brew of standards, with more to come.

In the following section is a brief survey of the work of DAVIC, IEEE 802.14, MCNS, and IETF IPCDN.

Digital Audio Visual Council (DAVIC)

DAVIC was the first organization to complete a MAC/PHY specification for HFC networks. The original specification is DAVIC 1.1, ratified in September 1996. Because their principle work is targeted toward digital video to the home, they made heavy use of Digital Video Broadcasters (DVB) specifications, which in turn made heavy use of MPEG-2 and ATM. The DAVIC HFC specifications are available at `http://www.davic.org`.

Key features of the specifications are the following:

- QAM-64 modulation on the forward channel with fixed interleaving using ITU J83 Annex A. (See Table 2–3 Annex A and Annex B comparison.)

- MPEG-2 TS/ MPEG-2 coding.

- QPSK modulation on the return path with no interleaving; only CRC error checking.

- Support for a variety of Media Access Control techniques. DAVIC 1.1 supports contention, reservation, and time division multiple access modes.

- ATM framing for out-of-band signaling.

- Use of an out-of-band signaling channel in the forward path.

IEEE 802.14

The IEEE 802.14 committee (http://www.walkingdog.com) began studying PHY/MAC issues of data/cable in November 1994.

The charter of the committee, quoting from their web page, read as follows:

> "The IEEE 802.14 Working Group is chartered to create standards for data transport over traditional cable TV networks. The reference architecture specifies a hybrid fiber/coax plant with an 80 kilometer radius from the head end. The primary thrust of the network protocol in design is to transport IEEE 802.2 LLC traffic types (exemplified by Ethernet). There is, however, a strong feeling within the group that the network should also support ATM networking to carry various types of multimedia traffic."

The scope of their work is limited to the MAC/PHY and depicted in Figure 3–6.

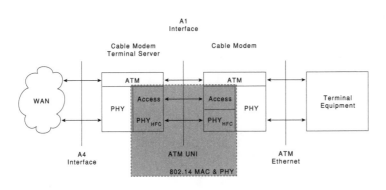

Figure 3–6
IEEE 802.14 scope of work.
Courtesy of IEEE
(http://www.walkingdog.com)

The committee is currently in the consensus-building phase, with many decisions already made.

As of this writing, the key features of the specification are as follows:

- QAM-64 modulation on the forward channel with fixed and variable interleaving using ITU J83 Annex A. (See Table 2–3 Annex A and Annex B comparison.)

- QPSK and QAM-16 modulation on the return path with no interleaving; only CRC error checking.

- ATM framing mandatory on the forward and return paths.

- Variable length, packet mode framing on the return path, considered as an option.

- Use of minislots for grant requests.

- Use of ternary tree backoff for collision resolution.

- Diffie Hellman key exchange.

Multimedia Cable Network System Partners Ltd. (MCNS)

Partly owing to the slow pace of IEEE 802.14 development, a group of North American cable operators chose to take a short-cut to standardization. MCNS (http://www.cablemodem.com) is a partnership among several North American MSOs to standardize physical-layer, MAC, and security protocols to provide data services over HFC.

The basic specifications include the following:

- The radio frequency (RF) interface

- The network security specification

- The operations support system interface

It was not a goal of these specifications to incur excessive risk or delays by pushing the limits of technology and performance. The key criterion was time to market.

More specifically, their goal is to meet basic feature and cost requirements for a product designed to last 3–5 years. The basic feature requirement means support for the IP protocol and for multiple grades of service. The cost requirement means the first-generation product will retail for under $400 with second-generation product retailing below $250.

The specification would be based on technology that has been implemented and tested as of the specification date, April 1997, and that has enjoyed actual deployment. This gives prominent roles to Broadcom, General Instruments, and Bay Networks (LanCity). Broadcom has a strong position among semiconductor providers in providing QAM and QPSK modulators and demodulators; GI has extensive experience with cable systems; and Bay/LanCity has a functioning MAC.

Although the initial standard was based on available technologies, the protocols are layered so as to be decoupled. Language in the specification exists to support future upgrades. The product is expected to be available by the end of 1997.

MCNS consists of four firms: Comcast Cable Communications, Inc., Cox Communications, Tele-Communications, Inc., and Time Warner Cable. MCNS in turn has partnered with Rogers Cablesystem (Canada), Continental Cablevision (now Media-One), and Cable Television Laboratories (CableLabs), to participate in, and support, this project.

MCNS released a preliminary specification in December 1996 to 95 industry vendors for comment. Comments were received in January 1997. A final specification was released at the National Cable TV Association convention in April 1997.

As of this writing, key features of the specification are as follows:

- QAM-64 modulation on the forward channel with variable interleaving using ITU J83 Annex B. (See Table 2–3 Annex A and Annex B comparison).

- QPSK modulation on the return path with no interleaving; only CRC error checking.

- MPEG-2 Transport Stream framing on the forward channel. Data will be encapsulated in MPEG TS frames.

- Variable length, packet mode framing, with ATM cells as an option.

- Grant requests sent in minislots that contain the amount of bandwidth requested.

- Use of binary exponential backoff for collision resolution.

- Full SNMP MIB required in the cable modem, requiring an IP address for the subscriber modem as well as the subscriber computer.

- RSA encryption.

In addition to technical specification, MCNS added provisions for a licensing pool. To provide an incentive for vendors to submit intellectual property to the partnership, MCNS offered to collect royalties from vendors that manufacture according to the specification. MCNS would then pass royalties to the original contributors.

Many vendors announced immediate support for the specification, even though their current products do not conform. The list included 3Com, Bay Networks (LanCity, a prominent participant in MCNS discussions), Cisco Systems, Com21, General Instruments, Hewlett-Packard, Hybrid Networks, Intel, Motorola, NEC, Panasonic, Scientific Atlanta, Sharp Electronics, Terayon, Toshiba, US Robotics (since acquired by 3Com), and Zenith Electronics. Other supporters include chip suppliers, the most prominent of which is Broadcom (Tustin, California), which developed early versions of QAM and QPSK modulator/demodulator ASICs.

Table 3–3 compares the standards recommended by DAVIC, IEEE, and MCNS.

Table 3–3   *DAVIC, IEEE 802.14, and MCNS Specifications*

	DAVIC	**802.14**	**MCNS**
Ratification date	September 1996	N/A	Still in April 1997 draft form
Downstream modulation	QAM 16 QAM 64	QAM 64	QAM 64 QAM 256
Downstream FEC	Annex A	Annex A and B	Annex B
Upstream modulation	QPSK	QPSK, QAM 16	QPSK, QAM 16
Upstream bit rates	0.256 1.544 Mbps 3.088 Mbps	0.512 Mbps through 8.192 Mbps	0.256 Mbps through 8.192 Mbps
Control channels	Out-of-band and in-band	In-band	In-band
Framing	ATM	ATM	MPEG
Key distribution	Vendor specific	Diffie Helman	RSA

IETF IP over Cable Data Networks

The Internet Engineering Task Force has convened a working group called *IP over Cable Data Networks (IPCDN)*. Its charter addresses service management issues related to the transport of IP of cable networks. It is planning to build on the work of MCNS and 802.14 and is not specifying a PHY or MAC protocol. The objectives are to specify operations for IP over cable networks. This involves the development of reference models and defines support for IP services such as:

- IP address acquisition by cable modems, using the Dynamic Host Configuration Protocol (DHCP)

- Bandwidth guarantees for IP, using the Resource Reservation Protocol (RSVP)

- Security, using the IP security specification (IPSEC)

- Management information, using the Simple Network Management Protocol (SNMP)

DHCP, RSVP, IPSEC, and SNMP are existing Internet recommendations. IPCDN is to specify what modifications, if any, need to be made to these to operate in the cable environment. To fulfill this mandate, certain deliverables are required.

IPCDN's earliest deliverable is "Definitions of Managed Objects for HFC RF Spectrum Management," by Masuma Ahmed of Terayon and Mario P. Vecchi of Time Warner Roadrunner, which specifies management objects to be contained in a new cable modem MIB. This document can be found at the Internet drafts archive at:

`http://www.isi.edu`

in a document titled `draft-ahmed-asmimib-mib-02.txt`.

Other deliverables include:

- **Architecture RFC.** Essentially a reference model that provides a common vocabulary for components and interfaces

- **IP over CATV RFC.** Deals with CATV-specific issues, including spectrum/bandwidth management, framing and encapsulation, and security.

- **CATV network requirements.** RFC that includes language on MIBs and RF management.

As of this writing, resolution of these issues will be drafted in December 1997.

COMPARISON OF VIDEO AND DATA SUBSCRIBER UNITS

One of the key differences between video and data is the tradeoff between latency and high integrity bit rate. In comparison with video, data can withstand a relatively high bit error rate. Data has other error correction mechanisms, whereas bad bits in MPEG flows quickly degrade picture quality. Therefore, video uses substantially more interleaving, generating substantially more latency. This has lead to a difference in standardization.

In addition, the basic service for video will continue to be carried in an MPEG-2 TS format because:

- Satellite distribution to the head ends is in MPEG TS format.

- Efficiency is paramount when you are carrying hundreds of compressed channels that equate to several hundred Mb. Ten percent overhead would increase spectrum requirement by several 6 MHz channels. MPEG has the least overhead.

- Security is fundamentally different in broadcast networks.

Apart from interleaving and the use of MPEG, there are many common hardware and software elements. Some of the commonalties are as follows:

- Use of QAM modulation on the forward path
- Use of QPSK modulation on the return path
- Use of concatenated Reed-Solomon forward error correction
- The requirement for a MAC protocol for the return path
- The requirement for local processing to manipulate digital content

The convergence of digital TV set tops with cable modems is encouraged by the fact that set tops are likely to have a built-in IP stack. Digital video content providers are providing access to web content to augment their programming. This will require an IP stack and programming languages such as JavaScript. IP will also be used to signal video servers and provide the network plumbing for electronic program guides. Ethernet controllers are very inexpensive, so the incremental cost of having an Ethernet port for pure data purposes inside a digital set top is minimal. Putting IP into a digital set top will eventually lead to little distinction between a cable modem and video decoder. OpenTV (by Thomson Sun Interactive), PowerTV (by PowerTV of Cupertino, California), and David by Microware (Des Moines, Iowa) are three operating systems for digital TV set tops that have built-in TCP/IP stacks.

The convergence of TV and cable modems suggests that internet-working vendors that build routers for the Internet and consumer electronics vendors that build televisions and set tops will converge, each industry will need to acquire the skills of the other.

BUSINESS OBSTACLES TO RBB

Despite cable's lead in residential broadband, 1996 was a financially bad year for cable operators. With all the developments of digital TV and Internet for the masses, the financial situation of cable operators and their equipment suppliers, along with their share prices, languished amidst one of the strongest stock markets in history.

Upgrade Costs

Cable operators are faced with a large bill for HFC, two-way and data service upgrades yet to come. The cost components are HFC upgrade, two-way upgrades, head-end data equipment, head-end digital video equipment, and customer premises equipment. An estimate of capital cost per subscriber is as follows:

HFC upgrade ($20 K/mile; 80 homes passed/mile; 62.5% take rate)	$400
Two-way upgrade	$50
Head-end equipment (CMTS, video and web servers, management systems)	$250
Subscriber equipment	$250
Test equipment, tools, training, initial marketing	$50
Total Costs	$1,000

Using interest rates of 9 percent and straight line depreciation over 15 years, normal for the cable industry, the interest and depreciation on the incremental capital is calculated as:

```
(0.09 + 0.066) * 1,000
```

or $156 per year or $13 per month

This is a substantial capital investment for an industry with over $50 billion (over $900 per subscriber) of current debt and whose debt to cash flow ratio exceeds 5.5.

The Cost of Digital TV

Despite the fact that analog TV does not afford as many additional channels as digital, MSOs may elect to upgrade to analog instead of digital because of cost. Going digital would add an incremental $200 per set top above the cost of an advanced analog set top and a $3–$5 million investment at the head end to capture, convert, edit, and store local content in digital format and support ad insertion from analog source materials. For a system serving 100,000 subscribers, the incremental cost of digital is on the order of $25 million.

Improving analog TV would provide up to 550 MHz of forward bandwidth for up to 80 channels and would add new features such as electronic program guides. MSOs in the United States have ordered over four million new, advanced, analog set tops through 1997.

An operator could deploy the cheaper advanced analog TV service and risk losing a percentage of his subscribers to DBS and wireless cable or spend the millions and retain the same subscribership. Depending on customer erosion, determined by customer satisfaction with advanced analog, the operators might stay with analog. This strategy works if DBS sales level off and MSOs are able to repair their balance sheets over the next few years.

Competition

DBS is currently the greatest competitive problem for the cable industry for three reasons. The first is the possibility that cable subscribers will disconnect and subscribe to DBS instead. This occurred after cable rate increases in 1996, and after the Telecom Act of 1996 passed. Interestingly, a majority of DBS subscribers continue to subscribe to cable to retain local programming. But the defections occurring after the 1996 rate increases show that there is limited room for price increases before consumers will choose DBS over cable.

A second competitive problem for cable is that DBS subscribers tend to be high-income households. In a national survey, the *Los Angeles Times* reported in December 1996 that for families with household income over $75,000 per year, cable take rates declined in calendar 1996 from 75.8 percent to 71.1 percent. High-income households are the ones that tend to opt for more expensive programming packages. When they turn to DBS for premium programming, they either cancel cable altogether, or scale back to the basic cable service. Currently, over 70 percent of cable subscribers accept some form of pay programming. If the percentage drops significantly, this would negatively affect the cash flows of the MSOs. In short, DBS gets the premium service subscribers, who generate better margins. (As a point of comparison, the average DBS subscriber spends about $52 per month, whereas the average cable subscriber spends about $35 per month.)

VIEWPOINT

Demographic Challenges of Cable

Cable penetration is low in Los Angeles anyway, about 62 percent, 3 to 4 percent below the national average. The reasons for this might be instructive. One problem is that the over-the-air reception is good and there are a lot of channels. The area is flat and the antenna farm on top of Mount Wilson provides good coverage for the 9.3 million residents of the L.A. Basin. More important perhaps is the multicultural nature of L.A. Relatively few cable programs cater to L.A.'s large Hispanic and Asian populations. Instead, these populations are served by a number of local over-the-air broadcasters who do not have access to national cable feeds. If there were more national ethnic programming, cable penetration among these populations might increase. But this is difficult to achieve with the current channel capacity crunch.

As a final competitive challenge, DBS has a relatively low churn rate. *Churn* refers to a customer's changing from one service or service provider to another. In many cases, DBS subscribers have prepaid for a year or more of service. Besides, if you pay a few hundred dollars for a dish, it is unlikely you will terminate service, even if you do not sign an annual programming agreement. At any rate, recent surveys conducted by Nielsen Media Research and Primestar indicate DBS subscribers tend to be very satisfied with their service.

In addition to competition from DBS is the prospect of digital TV offered by telephone companies or wireless cable operators. They have begun slowly, and the wireless cable operators are in poor financial health. But the technology they will use is largely the same as cable and satellite and there is significantly less cost to build the infrastructure because it is wireless.

Finally, there is the prospect of digital over-the-air broadcast. This has the potential to provide many more channels of free television. It remains to be seen what happens to programming and subscription television if there were 30 or more free digital channels per metropolitan area with a healthy measure of HDTV intermingled.

Market Saturation

Even without competition from satellite or telcos, cable take rates have not increased substantially since 1991, when they first topped 60 percent. Expected subscription has grown in 1997, and is expected to be less than two percent. It seems every TV household that wants cable has it. Cable's task is to get more revenue from households who already have it. This is the problem raised by DBS, with its big gains in pay-per-view and premium sports packaging and its superior penetration of high-income households.

Cable MSO management apparently agrees it is necessary to get more from each subscriber. Since the passage of Telecom 96, cable operators have taken the opportunity to raise subscription rates more than twice as fast as the consumer price index, clearly not a strategy for getting new households. Rates were raised an average of six to seven percent in the summer of 1996 with further increases in the first quarter of 1997. It remains to be seen how well the operators can continue to hold share in the face of these rate increases and increased competition.

Operational Issues

HFC brings significant improvements in field operations and network design. Even so, a wide area HFC plant is difficult to operate. The following are some recurring issues:

- **Amplifier Maintenance.** Adjust one amplifier and others need changing. Maintenance is also temperature dependent.

- **Power.** Fiber nodes and amplifiers in the field need electrical power. It can be supplied centrally or in a distributed manner. In either case, there are issues of battery maintenance, selecting proper voltage levels, safety, and corrosion. Power failures in cable systems are almost 10 times the rate as for telephone companies and are the largest single component of perceived video failures (see [Large].)

- **Topological redundancy.** This is expensive in a tree and branch topology. Generally, route diversity is not available between head ends and fiber nodes.

- **Remote network management.** Automated, remote management is generally not available for RF components. Troubleshooting is largely a manual process requiring experienced staff.

- **Staffing.** Upgrades are dependent on a pool of trained operations engineers. Cable wages for field engineers are roughly 10–20 percent less than unionized counterparts in the telephone industry. Unions have had little success in organizing trade labor in the cable industry, but this may change with recent cutbacks at some MSOs.

See Table 3–4 for failure rates of some system components [Large].

Table 3–4   *Failure Rates for Cable System Components*

Component	Annual Failure Rate (%)
Fiber-optic cable	0.06%/mile
Coaxial cable	0.12%/mile
Drop cable	1.5%/drop
Power supplies	5%

Table 3–4 *Failure Rates for Cable System Components, Continued*

Component	Annual Failure Rate (%)
Amplifiers	3%
Fiber transmitters and nodes	7%

Regulatory Issues

Cable franchise fees levied by cities produce about five percent of gross revenue. Whether or not this franchise fee would apply to data services is not clear; but for the moment, data and cable operators seem to be paying it. In addition to franchise fees, some cities levy their charges based on the amount of wiring done in the city. In Troy, Michigan, the home of an early and contentious legal argument over municipal franchise fees, the initial charges are 25 cents per foot per year for overhead wiring and 40 cents for underground. These charges are being applied to TCI for cable operations. As of this writing, this matter is going to court. Note that cable operators do not pay municipal franchise fees for telephone service because these franchises are established by agreement with the state public utility commissions.

Language in the Telecom Act of 1996 precludes excessive regulation and charging of fees by municipalities, with the view that such impositions would illegally restrict competition. Therefore, the Act preempts certain state and municipal rights. Whether or not the Troy, Michigan ordinance is unnecessarily restrictive is to be determined in court. Cities argue that they are entitled to compensation for the digging occurring on municipal property. Digging shortens the useful life of streets, and having wiring underground complicates and slows the maintenance process for other city services, such as sewage, which uses conduits close to those used by telecommunications operators.

Must Carry rules are a particular problem for the cable industry. The 1992 Cable Act required cable operators to carry local broadcasts, that is, local over-the-air stations. In part, this Cable Act was enacted to protect the viability of small stations that were threatened by the possible loss of over half their available audience if they were not on cable. Note that because DBS does not carry local broadcast stations, Must Carry rules do not apply to it. When DBS does begin carrying local broadcast, there will be regulatory and legislative pressure, as a point of equal treatment, for it to abide by Must Carry rules.

Because their analog cable networks are strained for channel capacity, cable operators view Must Carry rules as an economic burden. Many want to drop locally produced governmental, public access, educational, and religious programming in favor of, say, more shopping or pay-per-view channels.

The cable industry attacked the provision in federal court, and the case eventually went to the U.S. Supreme Court. The Supreme Court had jurisdiction because the case is framed by the cable operators as a First Amendment, Free Speech case. The cable operators argued that being forced to carry stations is an unconstitutional abridgment of their free speech rights. In a five to four decision handed down in April 1997, however, the court ruled against the cable operators, stating that the imposition on local broadcasters, who would be forced out of business, was greater than the imposition on the cable operator.

Must Carry rules are a burden in an analog environment, whether over-the-air or cable. Analog TV consumes too much bandwidth per channel, which limits channel capacity.

Cable, like over-the-air broadcast, will be subject to increased responsibilities for closed-captioning and emergency broadcasts. There is a requirement that all content be closed-captioned by the end of 2005. Cable operators estimate this will cost $500 million or more.

Until now, cable has been exempt from participating in the nation's emergency broadcast system. This changes in July of 1997 and will present new operating problems to cable operators. Requirements to participate in local and state emergency notification services are also possible. These services are suprisingly complicated and will require millions of dollars to implement.

Relationships with Programmers

Programmers and cable operators certainly need each other. But, their relationship is sensitive because of three problems, namely:

- Access to shelf space

- Access to basic service

- Programming fees

Access to Shelf Space

Until digital TV arrives, cable will be under pressure to provide shelf space for the dozens of networks wanting distribution. TCI recently announced it is dropping some channels, such as the WGN superstation from Chicago, in some systems and replacing them with the Learning Channel, Animal Planet, Home and Garden, and the Cartoon Network. The displaced programmers complain that the new channels are all owned by Liberty Media or Time Warner. TCI owns stock in both Liberty Media and Time Warner. FCC rules only require that programs owned by cable operators occupy no more than 40 percent of channel capacity; although TCI did not violate this rule, some programmers are unhappy.

Access to Basic Service

Even with digital TV, there is a need to manage the problem of channel glut. With digital TV, there is the possibility of 300 channels or more. But is likely there will be only 50–70 channels on basic cable because of navigation and audience fragmentation problems. All other channels need to be paid for and will therefore have less viewership. The decision of whether a particular network is or is not on basic is a sensitive matter between the operator and the network.

Programming Fees

Cable operators are growing weary of rising programming costs in general, but nowhere are they being hit harder than in sports, in which teams are looking to television rights to be their savior from escalating player salaries.

Cable operators pay 70–75 cents per subscriber per month for ESPN (owned by CapCities/ABC/Disney). But ESPN is proposing an increase to $1, to hang on to its half of the National Football League cable package it now shares with TNT.

In Los Angeles, sports programming has taken a new twist by the splitting of the Fox Sports Network into two channels, called Fox Sports 1 and Fox Sports 2. Fox acquired regional rights to broadcast live events for UCLA and USC football and basketball, Dodgers and Angels baseball, Lakers and Clippers basketball, and Kings and Mighty Ducks NHL hockey. This is a lot of regional sports, and Fox argued that they needed two channels because of time conflicts. So, Fox split the original Fox Sports

West into two sports channels. Fox Sports 1 enjoys widespread coverage in Southern California, but there is little coverage of Fox Sports 2, which will carry games of the Dodgers and NHL Mighty Ducks.

How much does Fox want for Fox Sport 2? Seventy-five cents, just like Fox Sports 1 and ESPN. Note that the Golf Channel, Classic Sports, SpeedVision, and Outdoor Life are charging roughly 20 cents. The cable operators are resisting, and Fox Sports 2 is off the air in most areas.

SUMMARY

HFC, two-way upgrades, and the commitment to digital services (TV and data) solve operational problems of cable operators and expand their product offerings. Consumers get more channels, programmers get more shelf space, Internet service providers get the speed necessary to attract advertising, and cable operators get a lift from their current financial woes. For now, it is the only game in town for residential broadband. The image of the industry was given a lift recently by an investment made by Microsoft in Comcast. This might portend further investments by Microsoft or others who want to distribute content via cable.

Nonetheless, cable operators face commercial and technical problems before they can become the full-service networks they aspire to be. Cable is the target for others wanting in on residential broadband. It must continue to invest or let new and old competitors take marginal revenues. This is painful when cable's base business has a high capital overhead and its financial position is highly leveraged. Table 3–5 summarizes the issues facing those operators that want to make the transition of cable to RBB.

On the horizon are alternative technologies that approach the residential broadband market differently and could pose significant competitive pressures to cable. The next chapter covers one such alternative, digital subscriber loop access networks.

Table 3–5   *Challenges and Facilitators for Cable as an RBB Network*

Issue	Challenges	Facilitators
Media	Shared media return path is susceptible to ingress noise, particularly from noise sources inside the home.	Hybrid fiber coax reduces deep cascades and associated reliability and equalization problems. Offers more bandwidth than just coaxial.
Noise mitigation	A shared topology means that a single noise source can adversely affect a large number of households.	Sophisticated noise mitigation and problem isolation techniques are under development and are being investigated in standardization efforts.
Signaling	New methods are required to remotely control the subscriber unit for return path use.	Subscribers are always connected, there is no need for a call setup process for data or video, which reduces signaling requirements compared with telephone services.
Data handling capabilities	Requires implementation of two-way HFC or use of telephone return.	Natural topology for Internet data, high-speed forward and low-speed return closely maps to web access.
Standards	Multiple standardization efforts, some of them contradictory with one another, in process. Standards organizations are trying to balance system optimization with speed to market.	There is recognition of the need for standardization so that cable modems and digital set tops can be made available at retail.
Regulation	Cable is regulated by sometimes aggressive and incompatible municipal franchise agreements. Must Carry, closed-captioning, and emergency broadcast rules incur costs.	Some relief will be supplied by the Telecom Act of 1996, which limits state and local regulatory rights.

Table 3-5 *Challenges and Facilitators for Cable as an RBB Network, Continued*

Issue	Challenges	Facilitators
Market	There is perception of market saturation; cable passes nearly 100 percent of homes in the United States and elsewhere, but has difficulty increasing take rates, especially among high-income viewers.	The only game in town as of this date for residential broadband, multichannel service.
Business/ Financial	Potential skilled labor shortages. Need to improve relations with programmers. Cable operators are highly leveraged, and continued development of HFC, digital TV, and data services will require further capital.	Cable operators own many sources of programming. Cable is diversifying into DBS through Primestar.

REFERENCES

Articles

[Adams], Michael. "ATM and MPEG-2 in Cable TV Networks, Parts 1 and 2." *Communications Technology*, December 1995 and February 1996.

[CED January 1996] "The Issue: Signal Theft." *Communications Engineering & Design*, January 1996.

Colman, Price. "Nowhere to Go But Up." *Broadcasting and Cable*, December 9, 1996.

[Golmie], Masson, and Su Pieris. "Performance Evaluation of MAC Protocol Components for HFC Networks." *Broadband Access Systems*. Wai Sum Lai, Sam Jewell, Curtis Stiller Jr., Indra Widjaja, Dennis Karvelas, Editors, *Proceedings, SPIE 2917, 1996*, pages 120–130.

[Large], David. "Reliability Effects of Choice in Network Powering." *Communications Engineering & Design,* March 1996.

[Lyons], Gary. "Evolving from FSA to Passive Cable Networks." *Communications Engineering & Design*, September 1994.

"Passage of the Telecommunications Act of 1996." *Cable World*, March 17, 1997.

[Prodan], Richard, Majid Chelehmal, Tom Williams, and Craig Chamberlain. "Consideration for HFC return path design." *Communications Technology*, June 1996.

[Sandino], Esteban and Corinna Murphy. "High Speed Data Services and HFC Network Availability." *Communications Technology*, January 1997.

Internet Resources

http://www.academ.com/nanog/feb1996/data.over.cable.html
(Milo Medin's view of the world—he likes cable.)

http://www.aescon.com/cableonline
(Cable programming.)

http://www.cablelabs.com/
(CableLabs, the research arm of North American cable operators.)

http://www.cablemodem.com/
(The web home of the MCNS Partners.)

http://www.catv.org/
(Broadband Bob keeps you abreast of cable.)

http://www.civic.net/lgnet
(Local Government Network web page; occasionally discusses local broadband networking and regulatory matters.)

http://clover.macc.wisc.edu/
(Cable information from the webmaster of the Society of Cable TV Engineers.)

http://www.fcc.gov/csb.html
(Federal Communications Commission (FCC) cable activities.)

http://www.home.net/
(@Home is a system integrator for data services over cable networks.)

http://www.mediaone.com
http://www.cox.com
http://www.comcast.com/
(MediaOne [was Continental Cablevision], Cox Communications, and Comcast tell their stories.)

http://www.mint.net/sctv/
(State Cable TV home page.)

http://www.multichannel.com/
(A good source for Cable TV and telecom news.)

http://www.pathfinder.com/
(Time Warner's Pathfinder Service, a repository of their publications online.)

http://www.rogerswave.ca
(Rogers Cable in Toronto discuss their data and cable service called Wave.)

[Gillett] http://www.tns.lcs.mit.edu/publications/
(Sharon Eisner Gillett's master's thesis at MIT in which cable is cheaper than ISDN.)

http://www.corningfiber.com
(Corning fiber.)

http://wkgroup.com/ced/
(Communications engineering design.)

http://www.zdnet.com/~intweek/
(Interactive Week.)

CHAPTER 4

xDSL Access Networks

Cable operators offer their HFC networks for RBB services. How are the telephone companies to counter? Some telcos have toyed with the idea of building HFC networks of their own or buying cable operators. Among these are PacBell, Telecom Australia, and Ameritech. However, most telcos prefer to capitalize on their existing multibillion dollar embedded base of telephone wire by pushing the evolution of an existing technology called *Digital Subscriber Line (DSL)*. DSL is a commonly used technique to aggregate subscriber lines from multiple households onto a single link—often fiber—for the purpose of improved operations.

DSL is deployed by U.S. telephone operators to modernize their wiring to households. U.S. phone companies have enormous maintenance requirements for their 167 million wired connections to customers. Regulatory relief and business practice make more than $20 billion per year available for telco maintenance budgets. In the maintenance process, certain upgrades are permissible that have the effect of increasing the bit rate and lowering operating costs.

In addition to existing DSL upgrades, multiple enhancements to DSL technologies have been emerging. The various DSL technologies discussed in this chapter are High bit rate DSL (HDSL), Asymmetric DSL (ADSL), Symmetric DSL (SDSL), ISDN Digital Subscriber Line (IDSL), and Very high bit rate DSL (VDSL). Collectively, these are referred to as xDSL. Before considering xDSL technologies, a quick look at the history and architecture of the current telco services will be useful.

CURRENT TELCO SERVICES

Conventional wisdom proves phone wire is too restrictive for high-speed service. The conductive characteristics of phone wire, however, have the capacity to carry millions of bits per second. The reason they don't is that high and low pass filters are on every phone line that suppress signals above 4,000 Hz and below 400 Hz. Given these filters, the data-carrying capacity of phone wire is limited to 56 Kbps, using the latest generation of modems. Filters are used on other services as well, such as those shown in Table 4–1.

Table 4–1   *Bandwidth Comparisons for Telco Services*

Service	Upper Limit Bandwidth
Voice service	4,000 Hz
Alarm service	8,000 Hz
ISDN	80,000 Hz
T1 using 2B1Q	400,000 Hz
T1 using AMI	More than 800,000 Hz
ADSL	More than 1,000,000 Hz

These filters were installed as a bandwidth-limiting technique to enable economical frequency multiplexing over long distance lines. If users could have 20,000 Hz per phone line, which is the frequency response of a modern stereo system, bandwidth on the backbone would be reduced by a factor of six. In addition, as noted in Chapter 2, "Technical Foundations of Residential

Broadband," pushing that much frequency through wires creates problems of attenuation, distortion, and resistance that negatively impact the integrity of the signal.

Even so, new modulation, equalization, and error-control techniques make the phone lines capable of bit error rates greater than 6 Mbps. How much greater depends on the length of the phone wire between the subscriber and the telephone office, the physical condition of the wires (corrosion, insulation, bridged taps, crosstalk), and its thickness (the thicker the wire, the less the resistance). Table 4–2 shows the trade-off of distance and speed over 24 gauge phone wire (0.5 mm in diameter).

Table 4–2 *Distance Versus Speed Over 24 Gauge Phone Wire*

Phone Wire	Speed	Distance
DS1 (T1)	1.544 Mbps	18,000 feet
E1	2.048 Mbps	16,000 feet
DS2	6.312 Mbps	12,000 feet
E2	8.448 Mbps	9,000 feet
1/4 STS-1	12.960 Mbps	4,500 feet
1/2 STS-1	25.920 Mbps	3,000 feet
STS-1	51.840 Mbps	1,000 feet

Before examining how technology is pushing this envelope, it will be useful to overview two current telephone services: *Plain Old Telephone Services (POTS)* and *Integrated Services Digital Networks (ISDN)*.

Plain Old Telephone Services (POTS)

Telephone switching equipment, which establishes phone connections, is located in buildings called *central offices (CO)*. Customers are connected to the CO over thin-wire pairs, also referred to as *local loops*. These thin-wire pairs are segmented in lengths of 500 feet. Lengths are spliced together as needed to reach from the CO to the customer's home.

The first 500-foot segment from the CO, 26 gauge wire, is normally 0.41 mm in diameter. This is the thinnest type of phone wire. After the first segment, the phone wire is often of a thicker diameter, such as 24 gauge (0.50 mm) or the thickest—19 gauge (0.82 mm) wire. Because attenuation is inversely related to thickness, the 19 gauge wire is reserved for customers who are farther from the CO. Resistance in the wire is a function of temperature as well as wire thickness. At 70 degrees Fahrenheit, 26 gauge wire induces about 40 ohms of resistance per thousand feet; 19 gauge wire induces 10 ohms per thousand feet.

Local loops are bundled in large cables called *binder groups*. The job of the field technician is to cross connect a drop wire to your home and then to a phone wire in a binder group. Fifty phone wires to a binder group is a typical configuration near the subscriber. Some binder groups contain as few as 20 local loops, others as many as a few hundred. Feeder cables—roughly the first 9,000 feet coming out of a central office—can have hundreds or even thousands of pairs bundled together. These are referred to as binder groups as well. Figure 4–1 shows a simplified connection between a CO and customers.

Figure 4–1
Wiring schematic pre-DSL.

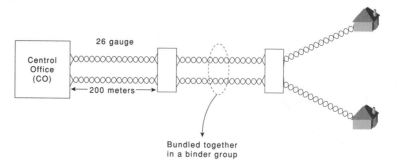

The copper in phone lines, although very thin, adds up when considered network-wide. U.S. telcos are the largest single consumer of copper in the world. The local loop consumes nearly 50 percent of telco capital cost. And capital cost of a local loop is $1,500–2,000 with about 50 percent being used for materials—including copper. Therefore, a premium is put on having the thinnest possible copper, consistent with transmission fidelity. Thinner wires mean less real estate and fewer material costs. Thinner wires are lighter, also reducing the cost of aerial runs. Generally, minimizing the consumption of copper is a goal in the design of phone systems, and one reason for telcos' interest in fiber technologies.

Eventually a single-wire pair is extracted from feeder cables and distribution cables of fewer wire pairs until the single-wire pair is connected to your house over the drop wire. The drop wire connects to your home in a junction box called the *network interface device (NID)*. The NID is typically a passive device that serves to couple external phone wire to internal phone wire in the home.

Basic Rate ISDN (BRI)

BRI is a digital service that provides 160 Kbps over phone wire up to 18,000 feet of 24 gauge wire. Its standard implementation (ANSI T1.601 or ITU I.431) employs echo cancellation to separate the transmit signal from the received signal. It uses bandwidth from 0 to about 80 KHz. (Some European systems use 120 KHz of bandwidth.) Therefore, provisioning of ISDN and analog POTS on the same local loop is not possible because both services utilize frequencies within the range of 400–4,000 Hz.

There were roughly 5.3 million BRI lines worldwide as of March 1997. In Germany and France, where success has been highest, approximately 5 percent of phone lines are provisioned with BRI.

Roughly 720,000 phone lines (about 0.6 percent) are provisioned in the United States. Estimates of the number of U.S. phone lines provisioned with BRI for the year 2000 vary from 2 million (Jupiter) to 4.5 million (IDC).

Limitations of Current Telco Networks

A number of limitations are associated with traditional telco networks that are currently driving interest in xDSL services.

Space and Distance Constraints

Currently, at least 22,000 local exchange carriers have their central offices in the United States, serving 167 million lines. Thousands more offices are served by long-distance carriers (AT&T, MCI, Sprint, and so on), also referred to as *interexchange carriers (IXC)*. The average CO serves 7,500 lines; a few serve as many as 100,000 lines. The problem of limited space in conduits in and out of central offices is becoming more severe as consumers add second lines for home use. About 20 percent of U.S. homes have added second lines for use in home offices and talkative teenagers.

Central offices are big, expensive buildings. Reducing real estate needs by using more compact electronics can account for noticeable savings. In places where real estate is truly at a premium, such as Rome and Tokyo, the sale of telco property can be large enough to fund digital build-outs of telephone services. In other words, telcos can trade buildings and real estate for new digital infrastructures. Therefore, an incentive exists to move to more compact facilities. One way to facilitate such a move is by installing fiber for new services and distributed switching systems.

Users obtain services by connecting to the CO. Their distance from the CO dictates the cost of providing the service and in some cases the type of service received.

There are two standardized range limits. *Revised resistance design rules (RRD)* limit distance to 18,000 feet for 24 gauge and 15,000 feet for 26 gauge. RRD rules are used for ISDN. Another range limit, *carrier serving area (CSA)* used for POTS service, limits phone service to 12,000 feet for 24 gauge and 9,000 feet for 26 gauge. Half of U.S. lines are within 9,000 of a CO or remote terminal; 80 percent are within 15,000 feet. (The RRD and CSA distances are estimates because the actual distance is a function of line quality. If a particular local loop has severe impairments, the distances might be smaller than these estimates.)

Internet Problems for POTS and ISDN

Internet service over POTS and ISDN lines is readily offered because of the ubiquity of phone service and the widespread availability of modems and voice switches provisioned for ISDN. The problem for telcos is that both POTS and ISDN use the CO voice switch, and data has a different usage profile from voice.

The major difference is that data sessions are longer than voice calls. A 1996 study conducted by US West (`http://www.cix.org/Current/ Governance/Regulatory/bell_studies/us/us3.html`) showed that voice calls on average are 5.64 minutes in length. Calls to Internet service providers on average are 32.47 minutes—six times the length of a voice call. Longer sessions might mean the number of ports on the switches is insufficient, causing busy signals. When busy signals become excessive, the telco is obliged to buy more switching equipment. At $500–1,000 per line, this is an expensive proposition. US West estimates it would need to provision 720,000 voice POTS lines to serve 120,000 Internet users. The economic and practical impacts of these estimates are significant, particularly when extrapolated to reflect the continuing growth of Internet usage. US West feels the issue of data service on voice switches should be addressed by the FCC.

The difficulties associated with this discrepancy in call length are particularly noticeable in areas of the country that have a lot of Internet dial-up activity, such as California. Because of the inordinate costs to their infrastructure caused by data sessions, Pacific Telesis has asked that Internet service providers be subject to access fees.

Telcos are searching for ways in which data traffic can be off-loaded from voice switches to specialized data communications equipment. xDSL services are a method to do this. This is the basic difference between ISDN and xDSL: ISDN goes through a voice switch; xDSL bypasses it. This makes rollout costs incremental for xDSL, whereas ISDN requires telcos to upgrade their voice switches (at a cost of up to $500,000 a pop) before ISDN can be offered. Therefore, xDSL can be considered a lower-cost platform for data service, even though it offers faster bit rates than ISDN.

FUNDAMENTALS OF DIGITAL SUBSCRIBER LINE

Clearly, the desire to reduce pressures on existing voice switches and avoid having to buy new ones are powerful motivators for DSL technology.

Other motivators also exist; the following section briefly overviews these motivators and the architecture of DSL. Subsequent sections look at specific flavors of DSL in more detail.

Benefits and Business Rationale

A more distributed architecture arises from the development of DSL technology. As the number of telephone wires grows, the binder groups become unwieldy. DSL was devised in the 1980s by Bellcore and AT&T to multiplex traffic on the wires in the binder group onto four wires, thereby saving lots of copper, which in turn means savings in weight and costs. DSL also provides improved operational characteristics due to reduced inventory of discrete lines.

Consider a neighborhood of 24 homes that are served by a common CO. Normally, 24 pairs of wires are connected from the homes to the CO. An alternative approach is to use DSL technology.

Instead of 24 pairs of wires backhauled from the homes to the CO, the 24 pairs are terminated in the neighborhood in a device called a *remote terminal*. From the remote terminal, 4 wire pairs carry traffic back to the CO. This saves 22 wire pairs between the CO and remote terminal. AT&T Bell Labs (now Lucent) improved on this with their SLC-96 product. SLC-96 served 96 homes over 5 T1s, (4 T1s active and 1 spare T1). SLC-96 reduced the need from 96 pairs to 10 pairs. The remote terminal can be located at pedestals, vaults, telephone poles, or office buildings. Figure 4–2 shows the major components of a DSL connection between a CO and homes.

Figure 4–2
DSL schematic.

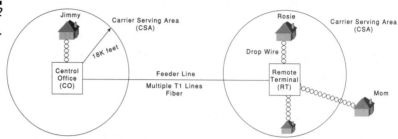

In Figure 4–2, Jimmy lives close to the CO, so his local loop directly connects to the CO. Rosie lives far from the CO, so she cannot be directly connected. Instead, her local loop connects to a remote terminal that is multiplexed back to the CO. Mom lives farther out still. She is connected to the remote terminal as well, but because she lives outside the range of the CSA she cannot receive advanced services, such as ISDN, without special equipment such as repeaters.

The connection between the CO and remote terminal can be fiber. Examples of such systems are the Lucent SLC-2000 and DSC Litespan 2010. Roughly 20 percent of lines in the United States use T1 or fiber DSL systems. These can be replaced by thousands of *controlled environmental vaults (CEV)*, which are underground sites or street-side pedestals that house new generations of remote terminals. In this architecture, the remote terminal can switch calls from Rosie to Mom.

By distributing switching functions, a telco can realize benefits in real estate. Another important savings for the telco is reduced electrical powering. Local powering for remote terminals can be used (with battery backup), which saves network powering—a big operational cost.

On a more strategic note, an important business rationale for DSL is that telcos require a competitive response to cable operators who want to offer full-service networking or enhanced data networking over HFC networks. Telcos have the option of building their own HFC networks or leveraging their existing phone wire plant, along with their points of presence in COs, CEVs, and pedestals.

Telcos' investment in existing plants is in the hundreds of billions of dollars with more than $20 billion spent annually on upgrades. Because they are enhancing their phone wires with new DSL technologies, which introduce fiber into the neighborhood, it makes business sense to investigate using fiber-based transmission coupled with phone wire to offer high-speed services.

Architecture

DSL networks are extensions of today's twisted-pair network offered by telephone companies. Types of DSL include IDSL (HDSL, ADSL, SDSL, and VDSL), Fiber to the Curb (FTTC), Fiber to the Building (FTTB), and Fiber to the Home (FTTH). The variations among these DSLs are distance, speed, and other factors derived from these, such as location of network termination and cost. The key differentiator is speed, which is determined largely by the distance from the household to where fiber is terminated.

Figure 4–3 depicts several types of DSL, including where fiber is terminated.

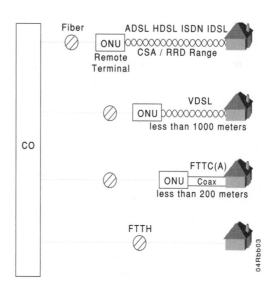

Figure 4–3
DSL flavors.

Fiber is terminated in the *Optical Network Unit (ONU)*, which is the equivalent of the fiber node in cable networks. Naturally, the farther you extend the fiber, the shorter the copper local loop can be and the higher the bit rate and the customer penetration rate can be. On the other hand, terminating fiber far into the network creates new problems of remote maintenance, remote powering, availability of remote real estate, up-front cost, and robustness of equipment.

The ONU may be situated as far into the neighborhood as the street pedestal (enabling either VDSL or FTTC), a multiple dwelling unit (enabling either VDSL or FTTB), or the individual residence (FTTH).

The following sections present several DSLs in more detail, beginning with ADSL (details of FTTC and FTTH are covered in Chapter 5, "FTTx Access Networks"). ADSL has received the greatest amount of hype to date. It has also received a fair amount of definition, whereas other DSL options are either proprietary or not yet completely defined. The vocabulary and specifications associated with ADSL provide a good foundation for discussing other DSLs.

ADSL (ASYMMETRIC DSL)

Originally conceived at Bellcore, ADSL is a telco-inspired service that offers a high-speed digital service and analog voice service over a local loop.

The original market driver was distribution of video on demand (VoD). When video on demand failed to be a hot market opportunity, the supporters of ADSL, like the supporters of HFC, latched onto high-speed Internet access for their business case. With or without video, people believed that more data would come downstream from the CO than upstream to the CO. The business case for ADSL rests strongly on the proposition that data and data alone, in this asymmetric form, is sufficient to drive infrastructure's development. This is yet to be determined in the marketplace, but is the same business proposition as data over cable.

A key difference between HFC and ADSL is that HFC is a shared medium and ADSL is not. An ADSL line is for the exclusive use of the subscriber, with no contention for bandwidth on that local loop. Therefore, despite the inherently higher bandwidth of cable, a crossover point exists where the ADSL service average bandwidth per user can exceed HFC.

Another important feature of ADSL is that it provides for passive transmission of analog voice service. This is a surprisingly complicated enhancement that the telcos are obliged to provide.

Two industry groups became involved in ADSL specification. ANSI T1.E1 took the responsibility for specifying a standard for modulation technique. This work culminated in document number ANSI T1.413, ratified in 1993. In addition, an industry consortium called the ADSL Forum (http://www.adsl.com) was created in 1994 to take the modulation techniques specified by ANSI and to discuss how to build end-to-end systems. The ADSL Forum is officially neutral on modulation techniques, although its members have strong opinions on the matter. The ADSL Forum is also specifying systems for a successor to ADSL called VDSL. Important dates in the evolution of ADSL are as follows:

- 1989: Pioneering work is published by Bellcore.

- 1993: Field trials begin using CAP modulation at British Telecom and Bell Atlantic.

  - ANSI T1E1.4 ratifies DMT as a modulation technique, based on the view that 6 MB video would be required for video on demand, specifically to carry four concurrent video streams.

- 1994: ADSL Forum is formed.

- 1996: CAP 1.5 Mb ADSL gets another shot after ADSL supporters agree video on demand is not the market driver it was expected to be.

- 1997: Trials go underway for CAP and DMT.

- 1998: Anticipated beginning of tariffed service.

ANSI specified DMT as the modulation technique for ADSL in T1.413. However, at that time work was being performed by corporate descendants of AT&T and the RBOCs using the well-known CAP modulation technique. These descendants were Paradyne and Globespan.

Despite the status accorded to DMT, these CAP advocates have continued to press on. The result is that ADSL comes in two flavors, depending on modulation scheme. CAP advocates have achieved some success with ANSI with the creation of an ad-hoc committee in late 1996 to investigate the standardization of CAP. As of this date, the possibility exists that a dual standard will be adopted.

Although differences exist concerning modulation techniques, end-to-end system architecture of ADSL systems is identical using either technique. That said, no discussion of ADSL would be complete without a comparison of CAP and DMT, the reason for one of the longest-running debates of ADSL.

CAP-Modulated ADSL

As of this writing, CAP-modulated ADSL is the more prevalent form of ADSL for use in trials. CAP is a well-understood technology, owing to its similarity with QAM. DMT is the official standard for now, and it has experienced many important commercial wins. But CAP supporters have not given up and are embarking on establishing CAP as a de facto standard, as well as continuing to press for standards status within ANSI. It remains to be seen if they can ramp up an embedded base before DMT gains momentum.

Globespan Technologies (formerly ATT Paradyne) is the licensing agent for CAP. As of this writing, 20 companies have been issued licenses. Among the licensees are Bellcore, Westell, Nokia (Finland), and NEC (Japan).

DMT-Modulated ADSL

Although CAP is well-understood and relatively inexpensive, some argue that it is difficult to scale because it is a single-carrier modulation technique and is susceptible to narrowband interference. DMT uses multiple carriers. At this point, DMT is capable of more speed than CAP. This is one reason ANSI committee T1E1.4 accorded it standards status in document T1.413.

This standard calls for 256 subbands of 4 KHz each, thereby occupying 1.024 GHz. Each subband can be modulated with QAM 64 for clean subbands, down to QPSK. If each of the subbands can support QAM-64 modulation, then the forward channel supports 6.1 Mbps. On the return path are 32 subbands, with a potential for 1.5 Mbps.

Among the providers of DMT systems are Alcatel, Amati, ECI Telecom, Westell, and Orckit. Alcatel caused a stir in the industry by winning a large procurement from several American telcos for a DMT-based service. Details of this win, the first important one for DMT, can be found on their web site at

```
http://www.alcatel.com
```

under the 1996 press release archives.

CAP and DMT Compared

CAP is a single-carrier technique that uses a wide passband. DMT is a multiple-carrier technique that uses many narrowband channels. The two have a number of engineering differences, even though, ultimately, they can offer similar service to the network layers discussed previously.

Adaptive Equalization

Adaptive equalizers are amplifiers that shape frequency response to compensate for attenuation and phase error. Adaptive equalization requires that the modems learn line characteristics and do so by sending probes and looking at the return signals. The equalizer then knows how it must amplify signals to get a nice flat frequency response. The greater the dynamic range, the more complex the equalization. ADSL requires 50 dB of dynamic range, making adaptive equalization complicated. Only with recent advances in digital signal processing (number crunching) has it become possible to have such equalization in relatively small packaging.

Adaptive equalization is required for CAP because noise characteristics vary significantly across the frequency passband. Adaptive equalization is not needed for DMT because noise characteristics do not vary across any given 4 KHz subband. A major issue in comparing DMT with CAP is determining the point at which the complexity of adaptive equalization surpasses the complexity of DMT's multiple Fourier transform calculations. This is determined by further implementation experience.

Power Consumption

Although DMT clearly scales to RBB and does not need adaptive equalization, other factors must be considered. First, with 256 channels, DMT has a disadvantage regarding power consumption (and therefore cost) when compared with CAP. DMT has a high peak to average power ratio because the multiple carriers can constructively interfere to yield a strong signal. DMT has higher computational requirements resulting in more transitors than the transceiver chips. Numbers are mostly proprietary at this point, but it is estimated that a single transceiver will consume 5W of power, even with further advances. Power consumption is important because hundreds or thousands (as carriers dearly hope) of transceivers might be at the central office, or CEV. This would require much more heat dissipation than CAP requires.

Latency

Another issue for DMT is that latencies are somewhat higher than CAP (15–25 ms versus 4–10 ms for maximum CSA). Because each subband uses only 4 KHz, no bit can travel faster than permitted by a QAM-64–modulated 4 KHz channel, about 24 Kbps. Throughput is greater, but latency is longer. The trade-off between throughput and latency is a historical one in data communications and has normally been settled in the marketplace.

Time to Market

At this point, CAP products have a time advantage of about a year over DMT products because of more integrated components and test equipment. However, this gap is expected to close rapidly, especially because DMT has enjoyed the standardization support of ANSI and ETSI and is being addressed by major chip vendors.

Speed

DMT appears to have the speed advantage over CAP. Because narrow carriers have relatively few equalization problems, more aggressive modulation techniques can be used on each channel. For CAP to achieve comparable bit rates, it might be necessary to use more bandwidth, far above 1 MHz. This creates new problems associated with high frequencies on wires and would reduce CAP's current advantage in power consumption.

Overview of CAP Versus DMT Summary

This discussion has tried to fairly represent the CAP/DMT debate. Without a doubt, advances will be made in both technologies, which will narrow the various technical gaps. Table 4–3 summarizes the important differences as they exist at the time of this writing.

Table 4–3 *Comparison of CAP and DMT for ADSL*

	CAP	DMT
Power consumption	Lower, fewer gates, currently twice CAP/line	Lower peak/average but will likely narrow gap power ratio
Forward carriers	1	256
Return carriers	1	32
Latency	4–10 ms	15–25 ms
Increment	320 Kb	32 Kb
Adaptive equalizers	Needed	None
Experience	More, including component integration and test equipment	
Licensing	Globespan	Amati, mainly
Maximum forward speed	E1	6 Mb
ANSI standardization	In process	Yes
Key competitors	Globespan, Paradyne, Westell	Amati (now Texas Instruments), Aware, ECI Telecom, Orckit, Westell

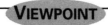

Convergence of CAP and DMT

In the end, the choice of modulation technique will likely be immaterial. DSP advances will enable these technologies to converge in cost and functionality and what we are then left with is the question of commercial interest.

The long-term challenges for DMT would appear to be power consumption and latency. In the end, the marketplace needs to speak loudly and clearly as to whether the speed of DMT is truly required by the subscriber.

ADSL Architecture

CAP and DMT advocates certainly differ on the merits of their respective modulation schemes, but they are coming together on further architectural elements of an ADSL system.

The ADSL Forum is neutral on modulation technique, leaving this discussion to ANSI. Instead, the ADSL Forum is intended to fill gaps left by ANSI with the objective of creating services from modem standardization.

The ADSL Forum produced a general schematic that is useful for establishing vocabulary. Their version is shown on the ADSL Forum web site at `http://www.adsl.com/`. A version of this schematic—modified slightly to account for DAVIC interfaces—is shown in Figure 4–4.

Content providers transmit information to the CO over the A9 interface (not shown) and the A4 interface that is seen in both Figure 4–4 and Figure 4–5. ADSL advocates anticipate that services will be received from digital broadcast, broadband networks, and narrowband networks, with network management information also terminated in the CO. The *DSL Access Multiplexer (DSLAM)* will house multiple *ATU-Cs (ADSL Transmission*

Unit—Central Office), as seen in Figure 4–5. The ATU-C is similar to a line card in the voice case; there is one ATU-C per subscriber. (This might change in the future, as discussed in the section "DSL Access Multiplexer (DSLAM)" later in this chapter.)

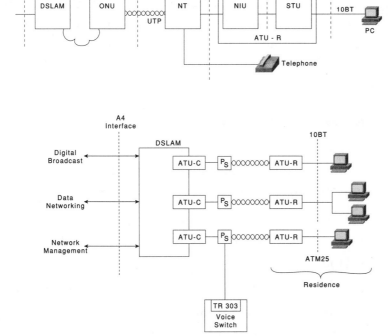

Figure 4–4

ADSL reference model, adapted from the ADSL Forum model.

Courtesy of ADSL Forum (http://www.adsl.com/)

Figure 4–5

More detail on an end-to-end design.

Courtesy of ADSL Forum (http://www.adsl.com/)

At the other end of the phone wire is an *ATU-R (ADSL Transmission Unit—Remote)* at the subscriber site. The ATU-R can support a multidrop or shared home topology. This means only one ATU-R is required even though multiple computers or televisions are connected through it within a home. The ATU-C and ATU-R engage in physical-layer negotiations between the home and CO. For the purpose of this discussion, they can be considered modems. The ATU-R will connect to terminal equipment in the home using either Ethernet or ATM.

POTS Splitter

The reference model also calls for a POTS splitter (labelled PS in Figure 4–5), which enables the carriage of analog telephone service and digital data/video service on the same line. A POTS splitter is a low pass filter that separates analog voice from ADSL frequencies. POTS splitting will be discussed in more detail in the section "Principles of Operation."

DSL Access Multiplexer (DSLAM)

The access node in the ADSL Forum reference model has acquired a new name as deliberation in the ADSL Forum has progressed. It is now referred to as the DSL Access Multiplexer, or DSLAM.

The primary functions of the DSLAM are to do the following:

- House a set of ATU-C interfaces.

- Statistically multiplex traffic from multiple ATU-Cs onto a single, high-speed trunk to the transport network—for example, over an ATM OC-3 circuit.

- Demultiplex traffic from the transport network and assign it to the correct ATU-C.

- Negotiate line speed (discussed later in the section "Rate Adaption").

- Serve as a central management platform.

- Cause ATM termination, in cases where packet mode traffic is delivered to the ATU-R.

- Select latency path, that is, the classification of traffic in the forward direction as either high-latency or low-latency traffic (discussed later in the section "Principles of Operation").

There is also discussion about where the DSLAM should be located. Certainly the DSLAM will be located in the CO. But should there be DSLAMs at the remote terminal or at the curb? Current product plans call for a single DSLAM to support 500–1,000 subscribers. If they have 6 Mb, then a 1,000-port DSLAM must be capable of forwarding 6 GB without congestion on the composite uplink. Bandwidth on this scale suggests that statistical multiplexing is required on the uplink.

Another possible enhancement to the DSLAM is to offer some form of port sharing. In the current model each ATU-C is dedicated to a corresponding ATU-R. Because the ATU-C is engineered as a DSP, it is expensive. Various vendors are considering how to overbook the ATU-Cs by having the *analog front end (AFE)* for each line terminate in a separate box. These ideas are proprietary and preliminary, but worth pursuing, because sharing of the ATU-C could substantially reduce the cost of the DSLAM. In this case a given user might not be able to connect to the ADSL network in the same way a POTS user sometimes receives a busy signal.

ATU-C and ATU-R Operations

The ATU-C and ATU-R are mostly concerned with the physical layer.

The following issues are involved:

- Frequency allocation

- Echo cancellation and Frequency Division Multiplexing (FDM)

- Rate adaption

Frequency Allocation

Bandwidth on the phone wire is divided into three parts; they are allocated for legacy analog POTS service, upstream digital service, and downstream digital service.

The POTS bandwidth allocation is well-known because it must conform with current use. POTS currently occupies 300–3,400 Hz. Therefore, this is not used by the ADSL digital services.

Bandwidth for digital services has two options. One is a frequency-division multiplexing mode, in which upstream and downstream bandwidth are in separate spectra. FDM simply has two half-duplex channels. Return path is transmitted in the range of 25–200 KHz. Forward path is transmitted from 250–1,000 KHz. The top graph of Figure 4–6 illustrates FDM frequency allocation.

As shown in the top graph of Figure 4–6, the lowest 3 KHz is used for POTS voice traffic. At 25 KHz up through approximately 200 KHz, upstream bandwidth is provided. Beginning at 250 KHz through 1,000 KHz, or even higher, downstream bandwidth is provided.

The issue with FDM mode is that the upstream and downstream channels need to be restricted so that they don't overlap. This reduces the amount of effective bandwidth these channels can use for modulation; therefore, greater speed is offered.

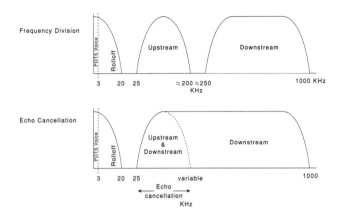

Figure 4–6
ADSL frequency allocation.

The alternative to frequency-division multiplexing is to extend the frequency range available to the upstream and downstream channels and thereby permit overlap bandwidths as illustrated in the lower graph of Figure 4–6. This case shows that the downstream bandwidth completely overlaps the upstream. In the range of overlap, echo cancellation is required.

Echo cancellation is used to obtain a clean signal in the event that both sides talk at the same time. Echo cancellation is used in V.32 and V.34 modems. It uses bandwidth more efficiently than FDM, but at the expense of complexity and cost.

In the echo cancellation mode, the sender transmits and remembers an analog signal. Milliseconds later, it detects a transmission from the other end. That transmission is a combination of real data from the other end and an echo of the original transmission. The sender extracts the echo from the combined reception and thereby can recover the intended transmission from the other side. Echo cancellation works well when only one echo exists, but if multiple echoes exist, primarily due to bridged taps, cancellation becomes more complicated and less effective.

Echo cancellation was invented for voice transmission over satellites. The long propagation delays (50 ms) annoyed listeners, who could hear themselves talk. Echo cancellation was created to alleviate this annoyance. Delays of less than 50 ms need not be cancelled because the human ear does not have that much time-sensitivity.

The problem for ADSL is slightly different from the voice/satellite case, but the same principles remain. A transmitter shares bandwidth with the sender on a full-duplex (fdx) channel. If ADSL is using a full-duplex channel, it must use echo cancellation to remove its interfering echo from the returned transmission. This is necessary because they are sharing the channel. With frequency-division multiplexing, echo cancellation is not needed.

A downside of echo cancellation is cost. It is slightly more expensive to have an echo canceller in the ATU-R than simply transmitting and receiving on different frequencies. Because of the desire for greater speed, however, echo cancellation seems to be gaining the upper hand. Moreover, with advances in signal processing, the cost penalty of echo cancellation is becoming less severe.

Rate Adaption

Another function of the ATU-C and ATU-R modems is rate adaption. *Rate adaption* is the process by which the modems determine the bit rate at which they will transmit. This is like the procedures used for data modems over analog lines. A negotiation process exists between the ATU-C and ATU-R to determine line characteristics (mainly impairments) and to agree on the maximum bit rate the local loop can reliably support for the duration of the session.

Because data sessions are long-lived, it is possible that line conditions might change during the session. For this reason, consideration is being given to periodically recalibrating the line to change bit rate on-the-fly. This yields complications for higher-level protocols such as ATM and MPEG and applications such as Real Audio, but is seen as key to providing maximum bit rate.

End-to-End Software Architecture

Many end-to-end software architectures are proposed for ADSL. The ADSL Forum is deliberating on a set of options in defining an end-to-end software approach. *System Network Architecture Group (SNAG)*, a working group within the ADSL Forum, is debating both ATM and packet-mode approaches. The architecture for packet-mode and ATM-mode delivery will be clarified by standards groups, but the choice of whether to have ATM to the home will be decided in the market.

Figure 4–7 depicts four end-to-end software options under discussion by ADSL. One option is to offer packet-mode service from the content provider to the terminal equipment. *Packet-mode service* refers to the native delivery of IP or MPEG Transport-Stream packets. The concern about this approach is the lack of bandwidth guarantees the network can provide. For this reason, many advocate some use of ATM.

It is widely believed that ATM will play a role in the core network. It is also widely believed that Ethernet will be used at the PC or end station, or that ATM will reside inside the PC. The question is where the Ethernet and ATM should converge.

Figure 4–7
End-to-end software
options.

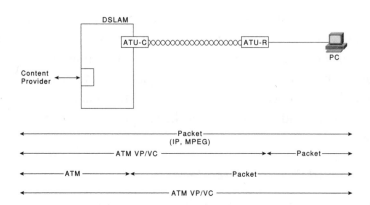

At the point where they meet, it is necessary to have a *segmentation and reassembly (SAR)* function. This is the process by which Ethernet packets are julienned into ATM cells and, conversely, ATM cells are reconstituted into Ethernet packets. Ethernet packets can be as large as 1,500 bytes and ATM cells as large as 48 bytes. Chopping the packets into cells is not a large problem. The bigger problem is how to associate the addresses inside Ethernet packets to ATM cell-addressing information, as well as how to enforce quality of service guarantees.

A second option beyond a pure packet approach is to have the SAR function in the ATU-R, also illustrated in Figure 4–7. Bandwidth guarantees are provided to the residence. This probably does not severely impact quality of service to the terminal equipment because congestion within the home is not a major issue. It does, however, put the responsibility for mapping ATM addresses to IP addresses in the ATU-R.

The third option shown in Figure 4–7 is to have SAR function in the DSLAM, where it can be performed in one place for multiple users. If the DSLAM has the SAR function, it will terminate a frame-based protocol from the ATU-R. Consideration is being given to ATM *Frame User Network Interface (FUNI)* and PPP.

In this model, the Point-to-Point Protocol (PPP) is used end-to-end, from the subscriber to the content provider. From the subscriber to the DSLAM, PPP packets are encapsulated in ATM FUNI packets. From the DSLAM to the content provider, PPP packets are encapsulated in ATM cells. The use of ATM FUNI is important because it permits ATM-style quality of service negotiation for packet mode delivery.

The final option shown in Figure 4–7 is to put the SAR function in the terminal equipment. This provides bandwidth guarantees to the subscriber, but possibly increases the cost of customer equipment. An approach using ATM to the PC is discussed at the Microsoft web site at `http://www.microsoft.com/supercomm/`. Microsoft has announced support for ATM in Windows and is supportive of the end-to-end ATM approach.

Connection Model

Another architectural issue is the connection model for ADSL. Will it use a terminal server model, in which a connection request is made by the end user when data service is requested, or will it use a permanently connected model, in which no connection request is needed?

The terminal server model is well understood by subscribers because it underlies the modem dial-up service used for online services today. A subscriber calls a phone number, an authentication server verifies the user's identity, and a connection is made. Potentially, it is possible for calls to get blocked. But given enough ports, connections are made reliably. The terminal server model has the advantage of port conservation. It is possible to provision ports for only two to five percent of potential callers on the proposition that not everyone calls simultaneously. As long as sessions are relatively short, significant economies of scale are achieved—as telephone companies and information service providers have known for years.

The problem with the terminal server model is that it is not compatible with push mode. When the Internet wants to send you something, you need to be connected, with a widely known IP address. It can be argued that push mode is crucial for scalability of RBB, as suggested in Chapter 1, "Market Drivers." Hence, momentum exists for the permanently connected model to be used.

The problem with permanent connections is that the subscriber is consuming network resources (address, port) when not actively using the service. The issue of connection model has yet to be decided by ADSL carriers.

Principles of Operation

Two services are provided by ADSL: transparent access to legacy voice service and high-speed digital service.

Voice Service

Given the frequency allocation of ADSL, the provision of voice service is relatively straightforward. ADSL service, mindful of the need to provide voice service, was designed to use frequencies above the voice band. POTS splitters used in the home shunt the 400–4,000 Hz frequencies to POTS wiring. Frequencies above this use whatever home wiring is provisioned for high-speed data service to get to the ATU-R. Care should be taken that the splitter will allow frequencies up to 20 KHz to pass to the NID. This enables alarm systems, which use frequencies higher than voice band, to continue functioning. ANSI T1.413 specifies 4 KHz, but most commercial filters will allow higher frequencies to pass. At the CO, a POTS splitter is also used to peel off the voice band and shunt it off to a voice switch.

POTS Splitter

The POTS splitter is a low-pass filter that shunts the voice traffic back to the NID, where it joins the in-home phone wiring. In the CO, a companion POTS splitter binds the customer phone line to a lead circuit in the telephone network.

There are three places at which the POTS splitter can be located in the home.

- Low-pass filter at the NID and the high-pass filter at the ATU-R

- Both filters in the ATU-R

- Both filters at the NID

When considering placement of the POTS filter, the following constraints are considered.

- The ATU-R requires local powering, so it might not be near the NID.

- Voice service should not be subject to impairments of home wiring, so it cannot be far from the NID.

- Low-pass filters are relatively large and will not fit into the NID.

These constraints suggest that the POTS splitter and ATU-R should be separated. If the NID, POTS splitter, and ATU-R are each in separate boxes, cost and distance are a concern. In particular, how far can the splitter be from the NID and the ATU-R, and how is a consumer to know where to best locate the splitter. One proposal from British Telecom is shown in Figure 4–8.

Figure 4–8
In-home configuration.

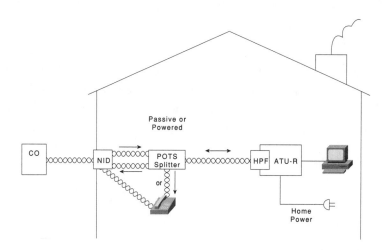

As shown in the figure, the local loop is terminated at the NID. Phone wire is used to connect the NID to the POTS splitter. The POTS splitter can either connect back to the NID for phone service, or it can connect directly to a phone. The latter case does not work when multiple phones are in the home. For this reason, wiring to support voice will likely make a U-turn and go back to the NID, at which point it will connect to in-home voice wiring. Wiring for high-speed data service continues to the ATU-R. The ATU-R might or might not have a high-pass filter associated with it. The ATU-R is powered locally and connects to the terminal equipment.

Although the exact location of the POTS splitter is unclear, it is clear that it will be on the customer's property, outside the purview of the phone company. It is the intent of most vendors that the POTS splitter be available at retail, with the telephone company charging an installation fee if required. Precisely how the output from the POTS splitter will connect to wires in the NID has not yet been finalized. Also, if the NID is located inside the home, as in the U.K., then protection against lightning is required, so home electronics won't be damaged. For these reasons, professional assistance is likely to be required for splitter installation.

Another issue associated with the POTS splitter is that it is an electromechanical device, so it won't come down in cost significantly.

Owing to the problem of POTS splitters, investigation has begun into a form of splitterless DSL, called *Consumer DSL (CDSL)*. It is also referred to as G.lite or ADSL Lite. It offers the services of ADSL but at a lower speed. At the time of this writing, technical feasibility was just underway.

Power

The ATU-R will be powered from the home. If power fails and the ADSL modems quit working, the telephone must continue to operate using network powering. A completely passive POTS splitter or an electrical relay in the POTS splitter will preserve telephone service in the event of a power loss.

Digital Service Startup

When the ATU-R is connected to the POTS splitter and local power, it will go through a start-up process to get connectivity at the physical and networking layer. Two processes are required. The first is to determine latency and the second is to determine bit rate.

Latency

Given that ADSL originally was intended for video transmission, the original designers planned for ADSL to be an error-free medium. To accomplish this, the bit stream was heavily interleaved and subjected to forward error correction. ANSI provided for 500 microsecond burst protection against impulse noise, which creates a 20 millisecond latency through the interleaver and modulators.

Data service has no such requirement for burst protection, because higher-layer protocols can perform error correction. The option of having no interleaving for data services exists, thereby reducing latency due to interleave delay. This low latency form has a modulator/demodulator delay of 4 milliseconds.

Therefore, ADSL provides for two logical data channels, a fast channel and a slow channel

Video traffic is directed through forward error correction and interleaving; hence, it is on a slow channel. Data traffic bypasses these functions; hence, it is on a fast channel. Bits from the two channels are multiplexed to form a composite stream that is then sent to line coding, or modulation.

How the ATU-C knows to funnel data to the fast or slow channel is negotiated at startup, but the mechanisms are an unanswered question at this time.

Bit Rate Negotiation

A neat twist on ADSL is *Rate ADaptive Subscriber Line (RADSL)*, which enhances CAP and DMT versions of ADSL by allowing the service to adapt to line conditions. The DMT rate can be adapted by using more or less aggressive modulation on each subband. CAP can be adapted by being more or less aggressive in its phase and amplitude modulation. RADSL is a negotiation process between the ATU-C and ATU-R by which they agree on the maximum bit rate sustainable as a function of the state of the line. Better lines result in more speed.

When the ATU-R is powered on, it probes the line to determine its signal-carrying capacity. Signal-carrying capacity is a function of distance (which means there is a ranging process, very much like that for HFC), noise (owing to bridged taps, crosstalk), and allowable margins. Negotiation can take different forms.

One form, adopted by ANSI, provides for four start-up rate options that are programmable under carrier network management control. Each option is a multiple of 32 Kbps. Thus, there is a large set of possible rates, but only four at a time can be offered by the ATU-C to the ATU-R. The ATU-R selects the highest bit rate possible consistent with its view of line conditions and allowable margins. It then informs the ATU-C of its choice, and clocks are set. If none of the four rates are acceptable, then the ATU-C can retry with the same set or a different set of four of the proposed rates.

Rate selection is communicated through network management to a *management information base (MIB)*. An ADSL MIB is yet to be defined, but the ADSL Forum has accepted the responsibility for defining it.

By offering a finite number of options, the service provider can offer services more predictable pricing. Other techniques would be for the ATU-R to offer a bit rate and the ATU-C to accept. Another approach is the Stalinist approach of not having any negotiation at all. One bit rate is offered, and if line conditions do not permit, then the call is blocked.

The entire process offered by ANSI is expected to take 12–20 seconds for the modems to agree on a bit rate. After the bit rate is negotiated, the ATU-R will cache line characteristics. If service is interrupted, then the ATU-R can use the speed previously negotiated to avoid the 20 seconds or so retraining period. It is expected that in a warm state the modem can recover from a service interruption in a second or two.

AT ISSUE

Signaling Line Condition to ATM

An interesting question of rate adaption is the relationship between the negotiated rate and ATM. ATM relies on knowing the speed to enforce quality of service. In particular, it manages two important variables: peak cell rate (the maximum number of ATM cells transmitted per unit time), and average cell rate. These are variables that are used to determine whether or not QoS service is available at the time the call request is made.

In an environment where the line speed is unknown until after session startup, or where the link speed changes during the life of the session, there needs to be a signaling mechanism whereby the conditions negotiated by RADSL can be communicated to ATM, so that ATM can enforce its QoS policies.

The signaling mechanisms to determine rate and latency path selection for RADSL and to communicate that information to ATM are ongoing issues. The result might be that RADSL options are few and ATM QoS guarantees are fairly coarse—that is, they are not sensitive to small changes in line state.

Autoconfiguration

Another task at startup is configuring of IP addresses and software filters, for the ATU-R. Because each subscriber has a dedicated ATU-C (port), the assignment of IP addressing and packet filters is relatively straightforward. They can be manually configured if the DSLAM is relatively intelligent. If not, some form of software tunnel is required to connect the ATU-C to a device with more intelligence.

During early stages of rollout, while there are a few hundred or perhaps a few thousand subscribers, manual software configuration is a possibility. When things get rolling, however, a new process might be called for, much like the cable industry is obliged to perform, whereby software configuration is done automatically.

The selection of an autoconfiguration technique is in part dependent on whether the ATU-R or the end system is IP-aware. If the ATU-R or the PC is IP-aware, for example, then the autoconfiguration technique can be a DHCP client, which requests an IP address from a DHCP server somewhere in the access network. This, of course, creates cost for the customer equipment. The implications of terminating ATM at various interfaces in the architecture are under investigation by the ADSL forum.

Digital Service Data Transfer

After the startup process is complete, data transfer is relatively straightforward. The ATU-R accepts data from the home network over an Ethernet, ATM25, or other standardized digital protocol. It encapsulates data using techniques mentioned previously in the section, "End-to-End Software Architecture" (refer to Figure 4–7), and forwards it. Unlike HFC, a media access control protocol is not required because the phone wire is not a shared medium.

Because data sessions can be long in duration, days long perhaps, line conditions could change during the life of the session. Increased temperature increases length and resistance of the wire, which are material both for phone wire and coaxial cable. *Dynamic rate adaption* refers to a process by which ATU-C and ATU-R renegotiate maximum bit rate during data transmission in response to changes revealed by periodic testing of line conditions.

Unfortunately, retraining the line during dynamic rate adaption will take as much time as static rate adaption, so a 20-second retraining period might be necessary while the ATU-C and ATU-R do their thing. During renegotiation, data will be lost.

It is undecided as to how frequently dynamic rate adaption should take place, if it is offered at all. Another unresolved question is whether retraining will be required on both the interleaved and noninterleaved traffic. Having retraining on both is complicated; retraining on data traffic (noninterleaved traffic) does not seem necessary.

Benefits and Challenges of Rate Adaption

Throughout the preceding discussion of digital service startup and data transfer, rate adaption is a recurring topic. It is an important technology to ADSL.

Certainly, rate adaption increases bit rate for customers with good phone wire. But the main interest carriers have in rate adaption is to expand their market coverage. Service providers are helped by rate adaption because they can cover greater distances; hence, they can sell to more potential users. Distant users get lower bit-rate service, but without RADSL they wouldn't get service at all. Rate adaption gives telcos comparable market reach to homes passed by cable.

Tests run in Australia determined that 65 percent of line work was for fixed rate E1. With RADSL, that ratio grew to 80 percent, although some distant users got less than E1.

A number of challenges, both technical and marketing, are associated with rate adaption.

When negotiating bit rate at startup, what metrics (characteristics to be measured), algorithms, and measurements will be used by the modems to agree to a bit rate? Possible metrics include distance, which requires a ranging mechanism, frequency response, signal to noise, and phase error. Converging to a set of metrics might be relatively straightforward; agreeing on the algorithms and measurements to derive bit rate is likely to be more contentious. Measurements and algorithms are typically not the sorts of things that can be standardized in public bodies. These are usually the province of competitive advantage and proprietary information.

Different companies will manufacture DSLAMs, ATU-Cs, and ATU-Rs. In particular, the ATU-R will be a retail item, manufactured by consumer electronics vendors who are not involved in making central office equipment. Therefore, agreement on measurements and algorithms is necessary for interoperability. Consumers should not need to concern themselves with purchasing a specific ATU-R that matches the particular ATU-C used in their neighborhood.

Such agreement will be hard to come by and might occur by pairwise collaboration among companies. That is, an ATU-C manufacturer and an ATU-R manufacturer will reach an agreement that will enable the particular combination of modems to be interoperable. Specifically, the metrics, algorithms, and measurements to calculate bit rate must be coordinated. Keep in mind that during renegotiation of bit rate, data will be lost. So, in addition to reaching agreement on how to calculate bit rate, vendors probably will need to agree on rules about when renegotiation is not required.

Another technical problem raised by rate adaption is how to synchronize information about the negotiated speed with information required at higher levels of networking. For example:

- How is a particular ATM virtual circuit mapped to a fast or slow channel? Is it by manual configuration?

- When the bit rate changes, should the network and the end system be informed, and if so, how?

- If bit rate increases during a session so that a traffic contract is violated, is it required that the ADSL go into renegotiation?

- If the connection is lost and a reconnection is made at a lower rate, what should the ATU-C or ATU-R tell ATM?

Besides the technical challenges, RADSL creates marketing and administrative problems for setting pricing and bandwidth guarantees. What do you charge the user when the user does not know in advance what speed he will get? Furthermore, the bit rate can change during a session as line conditions change. How this affects pricing is to be determined.

VDSL

Very high bit rate DSL (VDSL) pushes the envelope of what can be transmitted over 24 gauge copper pairs to the limit. VDSL's intent is to provide the fastest bit rate regardless of distance due to recent advances in DSP technology, which allow the implementation of sophisticated channel modulation and adaptive equalization schemes. The speeds planned by various flavors of VDSL are as follows:

12.96 Mbps	(1/4 STS-1)	4,500 feet of wire
25.82 Mbps	(1/2 STS-1)	3,000 feet of wire
51.84 Mbps	(STS-1)	1,000 feet of wire

Upstream rates under discussion fall into three general groups:

- 1.6–2.3 Mbps

- 19.2 Mbps

- Symmetric service at 1/4 STS and 1/2 STS

Multiple combinations of these speeds are under consideration.

Some operators, particularly in Europe, believe it is desirable to have both asymmetric and symmetric systems. Asymmetric systems would support 1.6 Mb/25.82 Mb and symmetric would support 12.96 Mb in both directions on shorter cables.

Agreement on speed in both directions requires agreement on a common spectrum allocation plan and is a prime issue for VDSL. Spectrum allocation planning will likely be discussed in ETSI TM6 and ANSI T1E1.4 as they consider modulation schemes for VDSL.

VDSL Architecture

The key to providing the fastest bit rate possible is to have the shortest possible length of copper wire and therefore the longest run of fiber possible. To this end, fiber is terminated in an ONU very near the subscriber. Current thought is that the distance from fiber termination to ONU is about 200 meters with an additional 100 meters on the customer premises, for a total distance of 300 meters from the fiber termination to the terminal equipment.

The ONU can be placed in many locations. If the subscriber resides within 200 meters of a CO or remote terminal, the ONU can be in the CO or remote terminal and VDSL service can be provided readily. For subscribers farther out, it will be necessary to locate the ONU in a street pedestal, controlled vault, or possibly a telephone pole. For residents of a multiple-dwelling unit, the ONU can be inside the development, but with any of these options, the potential penetration rate is low compared with longer reach services. It is unlikely VDSL will ever reach a majority of homes in the United States, although it may achieve higher penetration in Asia and Europe because of the higher population densities.

Comparison with ADSL

Because standardization is not imminent, we are limited at this point to comparing the intent of VDSL with ADSL. Certain advantages and problems of ADSL are noted and VDSL will address these in a subsequent generation.

POTS and ISDN Services

POTS and ISDN services will continue to be provided over phone wire, in the same manner as ADSL, using POTS splitters. Like ADSL, the VDSL subscriber unit will be powered from the home, but will relay to network powering in the event of a power outage.

VDSL Modulation Techniques

VDSL modulation techniques have yet to be standardized, but it is likely that early forms of VDSL will use well-known modulation such as QAM, CAP, and DMT. Consideration is also being given to *Pulse Amplitude Modulation (PAM)*.

Power Consumption

To reach the widest possible market, it is likely that VDSL termination will be located in places other than the protective and highly powered environment of the CO. Telephone poles and apartment basements might become VDSL sites. This puts a premium on heat dissipation, remote powering, and battery backup.

It is desired that each line card generate no more than 3W per channel and that there be some form of a "power down" facility when the line is not in use.

Dual Latency

It is generally agreed that VDSL systems should be designed for low latency and that the dual latency of ADSL should be avoided. This is because VDSL systems likely will be ATM-based, and avoiding the problem of dual latency on ATM is desirable.

Rate Adaption

Rate adaption is used by ADSL to extend the reach of the service and thereby improve its marketability. VDSL has no illusions about reach. It won't go farther than a kilometer in any case. Why bother about techniques to penetrate farther? Current thinking is to not allow rate adaption at startup. The bit rate will be set by the carrier and not by automatic negotiation between the subscriber modem and the carrier.

Spectrum Allocation

Early agreement on spectrum allocation is necessary. Without it, early generation VDSL products might become incompatible with succeeding generations. This is particularly important for VDSL, unlike ADSL, because VDSL occupies so much frequency.

The maximum passband probably will not exceed 30 MHz for 300 meter systems and 10 MHz for 1 kilometer systems. A key determinant of spectrum allocation is whether the same wire pair can support either asymmetric or symmetric service. If an operator decides to offer both services, asymmetric service will be slower to accommodate a common allocation scheme.

Comparison with Fiber to the Curb

VDSL has often been compared with Fiber to the Curb (FTTC). They have the same distance and speed characteristics, but VDSL uses phone wire for the last few hundred meters or so, whereas FTTC uses coaxial cable and is thus a shared medium. In fact, FTTC can be considered an extreme form of HFC. FTTC has facility sharing, a Media Access Control protocol, and asymmetrical service. Consider FTTC a form of HFC with the cluster being on the order of 10–20 users rather than the 500–2,000 of HFC systems.

The principal advantages of VDSL are the usual benefits of point-to-point systems compared to shared systems (MAC complexity, security) and the fact that replacing the phone wire with coaxial cable is unnecessary. On the other hand, the complexities of sharing are not great given the very small clusters of FTTC, and it remains to be seen how well phone wire will hold up at these speeds.

Standardization

Standardization of VDSL is undertaken by the ADSL Forum and the European Telecommunications Standards Institute (ETSI) Draft Technical Report on VDSL (DTR /TM-03068). A draft standard is due in 1998.

HDSL

ADSL and VDSL are the ambitious, high-speed forms of DSL. Additionally, both support analog telephony. Supporting both very high speed and voice results in complexity and cost, and many think that these services are not needed in a DSL system. What is required is a high-speed data service only, serving the needs of the burgeoning Internet market. No voice, no video.

Two intermediate, lower-speed, voiceless technologies are addressing this market. They are HDSL (as well as its derivative, HDSL-2) and IDSL.

Although ADSL enjoys considerable hype, HDSL is more widely deployed, with more than 500,000 lines deployed in North America today. The leading vendor of HDSL equipment, in terms of units shipped to date, is Pairgain Technologies (Tustin, CA). Other vendors are Adtran, ADC Telecommunications, and Orckit.

Originally developed by Bellcore, HDSL/T1/E1 technologies have been standardized by ANSI in the United States and by ETSI in Europe. The ANSI standard covers two-pair T1 transmission, with a data rate of 784 KBps on each twisted pair. ETSI standards exist both for a two-pair E1 system, with each pair carrying 1,168 KBps, and a three-pair E1 system with 784 KBps on each twisted pair.

HDSL became popular because it is a better way of provisioning T1 or E1 over twisted-pair copper lines than the long-used technique known as *Alternative Mark Inversion (AMI)*. HDSL uses less bandwidth and requires no repeaters up to CSA range. By using adaptive line equalization and 2B1Q modulation, HDSL transmits 1.544 Mbps or 2.048 Mbps in bandwidth ranging from 80–240 KHz over four wires. This is in contrast to the 1.5 MHz required by AMI. (AMI is still the encoding protocol used for the majority of T1.)

T1 service can be installed in a day for less than $1,000 by installing HDSL modems at each end of the line. Installation via AMI costs much more and takes more time because of the requirement to add repeaters between the subscriber and the CO. Depending on the length of the line, the cost to add repeaters for AMI could be up to $5,000 and could take more than a week.

HDSL is heavily used in cellular telephone buildouts. Traffic from the base station is backhauled to the CO using HDSL in more than 50 percent of installations. Currently, the vast majority of new T1 lines are provisioned with HDSL. But because of the embedded base of AMI, less than 30 percent of existing T1 lines are provisioned with HDSL.

HDSL does have drawbacks. First, no provision exists for analog voice because it uses the voice band. Second, ADSL achieves better speeds than HDSL because ADSL's asymmetry deliberately keeps the crosstalk at one end of the line or the other. Symmetric systems such as HDSL have crosstalk at both ends.

HDSL-2, or Single Line DSL (SDSL), is a slower but possibly more promising alternative to HDSL. The intention is to offer a symmetric service at T1 speeds using a single wire pair rather than two pairs. This will enable it to operate to a larger potential audience. It will require more aggressive modulation, shorter distances (about 10,000 feet), and better phone lines. Standards are set for 1998.

ISDN DIGITAL SUBSCRIBER LINE (IDSL)

IDSL is a cross between ISDN and xDSL. It is like ISDN in that it uses a single wire pair to transmit full-duplex data at 128 Kbps and at distances of up to RRD range. Like ISDN, IDSL uses a 2B1Q line code to enable transparent operation through the ISDN "U" interface. Finally, the user continues to use existing CPE (ISDN BRI terminal adapters, bridges, and routers) to make the CO connections.

The big difference is from the carrier's point of view. Unlike ISDN, ISDL does not connect through the voice switch. A new piece of data communications equipment terminates the ISDL connection and shunts it off to a router or data switch. This is a key feature because the overloading of central office voice switches by data users is a growing problem for telcos.

The limitations of ISDL are that the customer no longer has access to ISDN signaling or voice services. But for Internet service providers, who do not provide a public voice service, ISDL is an interesting way of using POTS dial service to offer higher-speed Internet access, targeting the embedded base of more than five million ISDN users as an initial market.

EARLY PROVISIONING FOR ADSL, HDSL, AND IDSL

Telephone companies are rolling out ADSL services cautiously. This fact has given pioneering Internet service providers an opportunity to offer some forms of DSL service in advance of the telcos' proposed ADSL services. Also, some corporations have considered implementing DSL services over private phone wire already installed in their buildings and industrial parks.

A popular underground method of obtaining DSL service is to order dry-pair wires from the telephone company and put ATU-C and ATU-R components on it. A *dry pair* is phone wire with no electronics from one end to the other. It may pass through a telco office, but it doesn't touch telco equipment. It is referred to by a variety of names by different telcos, including the following:

- Burglar alarm wire
- Local Area Data (LAD) circuit
- Modified metallic circuit with no ring down generators
- Voice grade 36 circuit

Controversy surrounds recent decisions by some LECs to terminate the availability of dry copper wires, resulting in frustrating attempts by some ISPs to offer early xDSL services. Public hearings in the Western United States, for example, are proceeding on petitions by US West to stop offering LAD service.

FACTORS IN CHOOSING WHICH DSL TO OFFER

When selecting which DSL to offer, the technical considerations that a carrier evaluates are length of drop cable, length of fiber, customer bit rate, number of homes supported per ONU, availability of real estate, and cost. They probably also should calculate real user demand for speed. Table 4–4 compares some of the characteristics that carriers must consider.

Table 4–4 *DSL Comparison*

	Modulation Scheme	Downstream Bit Rate	Upstream Bit Rate	Maximum Distance	Voice Support
ISDN	2B1Q	64, 128, 144 Kbps	64, 128, 144 Kbps	RRD	Active
IDSL	2B1Q	56, 64, 128, 144 Kbps	56, 64, 128, 144 Kbps	RRD	No
HDSL	2B1Q	2 Mbps	2 Mbps	CSA	No
SDSL	2B1Q	160 Kbps–1.168 Mbps	160 K–1.168 M	RSD 12 Kft	No
ADSL	CAP	1.5 Mbps–6 Mbps	64 K–800 K	RRD CSA	Passive
ADSL	DMT	1.5 Mbps–8 Mbps	64 K–800 K	RRD CSA	Passive
VDSL	TBD	12.96 Mbps–51.84 Mbps	1.5 Mbps–3 Mbps	1 kilometer 300 meters	Passive

TECHNICAL CHALLENGES TO xDSL

As discussed earlier in "Benefits and Business Rationale," telcos have a number of compelling business reasons to pursue xDSL. The benefit to the customer is faster service than POTS or ISDN. But although the general architecture of DSL service for POTS and ISDN is well-known, using the same infrastructure for high-speed services presents new technical problems.

Loop Qualification

The ideal local loop has no loading coils, no changes in wire gauge, no bridged taps, is well insulated, and is as short as possible. Reality is different. End-to-end performance of the local loop is impacted by loop length, loading coils, quality of end-to-end splicing of wiring segments, multiple changes of wire gauge, home wiring, age, corrosion, hostile binder groups, crosstalk, and bridged taps. Furthermore, most local loops are unnecessarily long as they snake around physical obstructions both indoors and outdoors.

Exactly how many of the America's 167 million lines or the world's 650 million lines are qualified to support ADSL? Telcos today are busy determining the answer using sampling, measurements, and experience gained from the rollout of ISDN and video on demand trials.

Crosstalk

Crosstalk, specifically *near end crosstalk (NEXT)*, is interference that occurs among wires wrapped in a single binder group. Each wire is unshielded; therefore, one wire pair can interfere with another through electromagnetic radiation. In Figure 4–9, Jimmy and Rosie are communicating with Grandma and Einstein, respectively. An example of crosstalk is when Rosie receives transmission from two sources. She hears intended traffic from Einstein and spurious traffic from Jimmy. She will not hear spurious traffic from Grandma because the distance will attenuate Grandma's signal. Hence, the problem is near end crosstalk rather than *far end crosstalk (FEXT)*.

Figure 4–9
Crosstalk visualized.

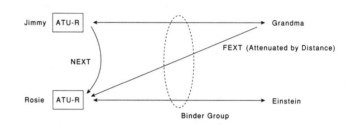

Rosie will: hear Einstein;
hear interference from Jimmy's high frequency transmissions;
not hear Grandma because high frequency transmissions
will be attenuated by distance

Crosstalk increases rapidly as a function of frequency. Because voice is a relatively low-frequency service, crosstalk is not a problem. But AMI-encoded T1 wires create a special problem because of the high frequencies used by AMI.

AMI requires repeaters 3000 feet from the central office and every 6000 feet thereafter, and consumes over 1.0 MHz of bandwidth. Hundreds of thousands of lines (T1 and E1) exist in the world today. But AMI encoding uses so much bandwidth, and creates so much crosstalk, that telephone companies cannot put more than one AMI-encoded line in a single 50 pair cable.

Because both ADSL and AMI T1 can use up to 1 MHz of bandwidth, they cannot coexist in the same binder group. Identifying which binder groups contain such T1 lines requires accurate record keeping, which has not always been maintained. Some carriers plan to restrict the copper plant that ADSL will use, keeping it separate from transmission technologies that could cause crosstalk.

ADSL uses frequencies up to 1 MHz. VDSL can go up to 10 MHz and can create a massive amount of crosstalk. Because the phone wires and binder groups to be used by VDSL were never specified for such high-frequency operation, crosstalk will be much higher than contemplated for modern cables.

Early deployment will not pose much of a problem because of the few number of users, owing to VDSL's short distance. But, as the network includes more and more VDSL systems, and crosstalk from other VDSL systems grows, the operator might have to replace some binder groups, resulting in a loss of service to the user as well as incremental installation costs.

One of the criticisms leveled at HFC is that shared media creates a loss of quality of service. In a sense, however, the same occurs for point-to-point service such as DSL. A binder group can be considered a shared media because wires in a binder group are not shielded from one another—users can affect other users.

BRIDGED TAPS

Bridged taps are stub extensions that go off to somewhere that a phone used to be. Bridged taps increase the chances of noise on the line because the unterminated tap can radiate power that reduces signal strength or becomes a source of ingress noise. In addition, the unterminated line can be a source of echo. Echo cancellation works very well when an echo has only one source, but when the original signal bounces off several places, cancellation becomes less effective. Bridged taps can be difficult and expensive to locate and fix.

Powering Remote Terminals

DSLAMs and ATU-Cs could possibly be located in controlled vaults and pedestals to increase neighborhood penetration and total available market. For VDSL, it might be necessary to locate equipment on telephone poles to justify a business case. These facilities have limited space, are out in the open, subject to vandalism, water damage, and other environmental problems, and require powering. One big problem is heat dissipation. xDSL services need to minimize power consumption so that a smaller amount of heat needs to be dissipated. For the moment, CAP has a 3–4W advantage over DMT in terms of power consumption. Local power will likely be used to power DSL equipment, but for emergencies, battery backup is needed. This also puts a premium on low-powered solutions.

Spectral Masking

ADSL is subject to multiple forms of long-term narrowband interference such as crosstalk. AM radio and amateur radio ingress are also available, both of which operate below 1 MHz. One way to avoid stationary narrowband interference is to mask out the offending frequencies and to avoid using them for ADSL transmission.

Such masking can be accomplished with notch filters, typically passive devices that notch out certain frequencies permanently. If the source of narrowband interference comes and goes, however, then passive filters might be insufficient. Active filters can handle such dynamic interference, but are more costly.

The term *spectral masking* refers to the process of notching out specific frequencies. In this environment, DMT operates rather naturally. Those frequencies that are notched out correspond to a certain subband and are not used by DMT. Because CAP modulates the entire spectrum, some adaptive equalization tricks are necessary to make it work with spectral masking.

Impulse Noise

Impulse noise on phone wire has been characterized by Bellcore and before that AT&T Bell Labs. It is fairly well understood. Duration is on the order of 100 microseconds with peak power around 10 millivolts. The solution is to use interleaving and Reed-Solomon. Current thinking is to use RS (200,216) coding for the interleaved (slow) channel.

Impulse noise is the phenomenon that precipitated the discussion on dual latency for ADSL and its complications for networking. One important source impulse noise is ringer noise from adjacent pairs.

ATU-R Maintenance

The ATU-R is more intelligent than the conventional POTS modem. For example, the ATU-R is capable of making software revisions, such as algorithms, for line calibration changes or if the carrier decides to change encapsulation protocols. Discussions are ongoing as to how software maintenance for the ATU-R is to be accomplished. Resistance is present to the carrier remotely managing it, in part because this is not the model used for current POTS or ISDN service. As in the case of cable modems, ATU-R maintenance covers new ground and can have serious cost consequences.

INDUSTRY CHALLENGES

In addition to technical problems, certain industry issues must be addressed before widespread rollout.

Provisioning Capabilities

Cable competitors are betting that telcos cannot provide and distribute xDSL services any better than they did ISDN. Given the still lengthy and confusing provisioning process for ISDN, this would seem to be a good bet. After all, only 5.3 million ISDN BRIs are installed worldwide a decade after product development.

Several significant issues must be overcome, such as provisioning, end-to-end protocol architecture, numbering plans, capacity planning, and workforce education.

Regulatory Issues

In the United States, the Telecommunications Act of 1996 required the local telephone companies to unbundle the local loop. This created potentially interesting consequences for ADSL service and telephone service.

As of this writing, it is not clear whether telcos are required to unbundle xDSL service. It is possible that only POTS service must be unbundled and that frequencies above it can still be used exclusively by the phone company. Of course, competitors could contravene the telcos' exclusive use of the higher frequencies by offering dry pairs to implement DSL service, but it is possible that this practice will be litigated.

Another unclear issue is what cost basis will be used to compensate the telcos for unbundling. The cost basis for POTS service provisioned through DSLs is not the same as that for direct connects: DSLs have lower costs (less trunking copper and smaller central office real estate requirements). States are currently calculating unbundled pricing for local loops, but it is not clear what cost model they will use to determine state mandated pricing. If states decide on a price based on wiring pairs extending from the home all the way to the CO, then the phone industry catches a break because costs for these are higher than costs for DSLs. On the other hand, if the state uses a cost basis that presumes a high percentage of DSL, then the telcos are not catching a break—they are not being compensated at a rate higher than actual costs. In response, they might extend DSL to reduce costs and meet profit goals implied by state mandated pricing.

Telcos are upset about unbundling for service reasons as well as financial ones. For example, consider the service issues when a subscriber of one of the new, competing companies uses dry pairs to obtain DSL service. Dry pairs can coexist on binder groups with normal rate-paying customers. It is possible for the user to inadvertently transmit excessively strong signals, thus creating sufficient near end crosstalk to impair the phone service of everyone else in the binder group. Telcos are worried about getting blamed from the other 49 voice subscribers. They are voicing such concerns as further arguments against unbundling.

Broadcast Video

Unlike HFC, xDSL, and even VDSL, are not competitive with broadcast digital TV. ADSL does not have the bandwidth and VDSL will likely have neither the bandwidth nor the coverage to compete with cable for video. The main use of video over DSL is for video on demand or near video on demand, neither of which have proven sufficient to justify massive infrastructure capital costs. Satellite broadcasting has the advantage for these services. The business case for xDSL primarily rests with data service.

If carriers adopt a terminal server model, they will be restricted to pull mode. If so, then the telcos are vulnerable to ISPs offering lower-speed DSL services over dry pairs and direct satellite providers providing the video.

Transitioning from Digital Services

Users transitioning from analog service to xDSL will be treated as new customers, but a transition plan needs to be put into place for users of existing digital services.

ISDN Transition

If a current user has ISDN service, uses both voice and data, and wants not to have his phone number changed, then certain tricks are needed to make this transition. This is an important class of user because they have already demonstrated the willingness to pay for a high-speed digital service, as well as with voice.

One approach is to replace the ISDN terminal adapter (TA) with an ATU-R and change the voice service from digital back to analog. This works as long as the protocols emitted from the ISDN TA and ATU-R are the same, like Ethernet. ISDN uses a dial model, whereas ADSL might use the always connected model. A few configuration changes might be necessary in the IP customer premises equipment and the home router to accommodate this change.

The more difficult case is if the user doesn't want to give up the ISDN and wants to add ADSL service on the same line. In this case, the ADSL spectrum can be located above the 80 KHz of IDSN and a splitter can be used to frequency multiplex the services on the same phone wire.

Relocated ADSL frequency above 80 KHz forces use of the echo cancellation mode and reduces bandwidth, primarily for the return path. Another approach is to embed ISDN data traffic within the ADSL traffic, using time-division multiplexing, for example, and forgoing the ISDN voice service.

If this is unpalatable to the user, another phone line can be provisioned.

Integration with Other Data Services

Many businesses and a few residences have access to other data services such as X.25 and frame relay. For example, telcos should consider how to provision these over ADSL as well, by using time-division multiplexing. They could just provision a new line, of course, but where a phone line already exists; in doing so they are losing some of the leverage of their existing infrastructure. It is in the interest of the telco to coexist with current services without provisioning extra physical circuits. Provisioning an extra circuit gives competing technologies an opening. It is undetermined whether and how to integrate xDSL with current digital services.

Home Networking

In addition to problems of external phone wiring, inside wiring can also introduce problems. Some estimate that in older buildings, roughly 10 percent of the home and office wiring and an additional 10 percent of the loop wiring is unusable for xDSL applications. Over time, home repairs, animal damage, corrosion, inside bridged taps, and withered insulation take their toll.

Bridge taps are the result of moving the endpoint of a wire pair from one location to another. When subscriber end #1 was moved to location #2, most of the time the telco left wire #1 in place, simply disconnecting it at the NID. This leftover wire stub typically isn't noticeable on a voice connection, but causes problems with high-speed connections (echo, reflection, and ingress for crosstalk and RF interference).

Fixing inside wiring will normally require a technician to come to the site and start over with new equipment. The main cost is technician time. Some telcos are fixing inside wire preemptively for xDSL trials to avoid maintenance calls later. One issue that concerns older homes is asbestos. Rewiring a home that contains hazardous materials is problematic for both the resident and the technician. For this reason, some consider wireless the best answer when asbestos is a known hazard.

AT ISSUE

Success Factors

Given the number of variables affecting the viability of xDSL, the following questions are among the most important for predicting its future:

- How many systems will be upgraded to DSL for POTS service? Using more DSL systems tends to reduce incremental cost for rolling out service.

- How many loops qualify? The higher the percentage of loops that qualify, including inside wiring, the higher the marketability of the service.

- How well can web servers deliver significantly faster service to showcase the speed of xDSL networks? If web servers or the Internet are chronically congested, the value of faster local loops is lost, the same as for cable.

- How popular will competitive technologies become, including ISDN and 56 KBps modems? These services share cross elasticity. Users who are content with 56 KBps or ISDN service might elect to stay with these.

- How will ISPs price high-speed Internet services?

- Is a market out there for access networks that can provide a full range of video and data services over a single network? If so,

> DSL will not qualify to satisfy it because DSL cannot offer broadcast video. How patient will telephone companies be if their services meet with initial consumer resistance?

SUMMARY

xDSL is the competitive response of telcos and some Internet service providers to cable operators offering high-speed data service. ADSL will be offered by telephone companies around the world. It leverages their existing wiring, an embedded investment on the order of a trillion dollars. It will also be offered by alternative providers such as ISPs, CAPs, and IXCs over nonservice carrying wire, known as dry copper or burglar alarm wire. The unbundling of local loops is a requirement of the Telecommunications Act of 1996, which affords alternative carriers access to dry copper. Table 4–5 summarizes the challenges and facilitators to xDSL's becoming a successful RBB network.

The next chapter addresses high-speed advancements beyond xDSL, including Fiber to the Curb (FTTC), Fiber to the Building (FTTB), and Fiber to the Home (FTTH).

Table 4–5 *Central Issues for Digital Subscriber Line and RBB*

Issue	Challenges	Facilitators
Media	Number of qualified existing loops is limited. Hostile binder groups, T1 encoded with AMI. Existing quality of home phone wiring is problematic in many cases.	Huge embedded base worldwide. Technical characteristics of the local loop permits higher speeds. DSL enhancements are underway for POTS service; reduces incremental costs for advanced DSL.
Modulation schemes	Competing standards—CAP and DMT—for ADSL.	CAP is relatively mature. DMT provides superior resistance to narrowband interference and can scale to high speeds.

Table 4–5   *Central Issues for Digital Subscriber Line and RBB, Continued*

Issue	Challenges	Facilitators
Noise mitigation	Multiple bridged taps create problems for echo cancellation.	Unlike cable, a point-to-point topology means a single noise source cannot adversely affect a large number of households.
Signaling	RADSL call setup is a time-consuming process. Mechanism required to signal latency path and RADSL to ATM. Mechanism required to signal dynamic rate adaption to ATM.	xDSL sessions can have a very long duration because they don't connect through a voice switch. This reduces signaling.
Data handling capabilities	Less speed than cable in the forward path for all but the most aggressive forms of ADSL.	Good return path service, enables reliable interactive service with known QoS. Multiple end-to-end software architectures.
Standards	Multiple standards at this point in time—modulation scheme, connection model, frequency allocation—that could inhibit interoperability of central office and consumer equipment.	Relatively few bodies are deliberating ADSL and VDSL specifications.
Business/Financial	Not a good track record at rolling out high-speed services. How to transition current ISDN customers to ADSL without rewiring. How to provide existing digital services, such as X.25 and frame relay.	Reduces telcos' real estate needs. Reduces data burden on voice switches and need for new ones. Leverages enormous investment in existing infrastructure. Maintenance budget can be used for certain upgrades.

Table 4–5 *Central Issues for Digital Subscriber Line and RBB, Continued*

Issue	Challenges	Facilitators
Regulation	Requirement to unbundle local loop inhibits telcos' incentive to invest in local loop technologies. Relationship to universal access unclear.	Telecommunications Act of 1996 tends to give ILECs a more free regulatory rein. Dry copper is available for competitive ISPs to provide service.
Market	Not suited for broadcast video. Not particularly well suited for VoD or nVoD.	Faster Internet service than POTS, 56 Kbps and ISDN service for consumers. The transition plan from HDSL through ADSL up to VDSL is logical.

REFERENCES

Articles

Allan, Dave (Northern Telecom). "Topics in DSLAM Architecture." *ADSL Forum 96-130*, December 1996.

Baines, Rupert. "Designing ADSL Devices." *Communications Systems Design Magazine*, November 1996.

Dugerdil, Bernard. "ADSL Access Network Reference Model." *ATM Forum 95-0610*, June 1995.

Kwok, Timothy. "Residential Broadband Internet Services and Applications Requirements." *IEEE Communications*, June 1997.

Kyees, Phillip, Ron McConnell, and Kamran Sistanizadeh. "ADSL: A New Twisted-Pair Access to the Information Highway." *IEEE Communications*, April 1995.

Maxwell, Kim. "ADSL, Transitional Technology for the Next 40 Years." *IEEE Communications*, November 1996.

Modlin, C.S. "ATM Over ADSL and VDSL: PMD Specific Attributes That Affect the Network." *ATM Forum 97-0156*, February 1997.

Rahmanian, Shahbaz (Ameritech). "Impact of the ADSL Rate Adaptation on QoS." *ATM Forum 96-115*, December 1996.

Rauschmayer, Kaycee, Dunn, Lin, and Stephens. "An End to End ADSL Service Model." *ADSL Forum 97-028*, March 1997.

Internet Resources

http://www.adsl.com
(The web home of the ADSL Forum; points to other ADSL sites as well.)

http://www.adsl.com/vdsl_tutorial.html
(VDSL tutorial from the ADSL Forum.)

http://www.alcatel.com/our_bus/telecom/products/broadband/whitep.html
(ADSL tutorial from Alcatel, early DSLAM providers.)

http://www.alumni.caltech.edu/~dank/isdn/adsl.html
(Dan Kegel's ADSL page.)

http://www.cix.org/Current/Governance/Regulatory/bell_studies/us/us3.html
(US West studies on Internet call length.)

http://www.cs.tut.fi/tlt/stuff/adsl/pt_adsl.html
(ADSL tutorial from the always innovative people at the Tampere Institute of Technology, Finland.)

http://www.globespan.net/cap_ppr.html
(Globespan discusses CAP modulation.)

Further information is obtainable in publicly filed documents with the Securities and Exchange Commission (SEC). Information from the prospectus filed by Orckit for their initial public offering in August 1996 was used in preparing this chapter. Other information is obtainable from the SEC for Amati, Aware, Westell, and other companies involved in xDSL products.

CHAPTER 5

FTTx Access Networks

Despite the advances in DSL technology, some still believe that that phone wire is not up to the job of residential broadband. One of the arguments is that phone wire was not designed for high-speed transport and even if it were, there are too many impairments owing to the age of the installed wires. In some parts of the country, the wire is decades old, which means corrosion and poor insulation exist. Certainly cable operators believe this, but some in the telco industry and various industry gurus think so as well. Therefore, consideration is being given to two alternatives that are intended to raise the bar on speed and reliability to and from the home for point-to-point wired networks. These networks are called *Fiber to the Curb (FTTC)* and *Fiber to the Home (FTTH)*. These terms are generic in that specific variations exist for both FTTC and FTTH. This chapter will overview the architecture and operation of both FTTC and FTTH, and the challenges and benefits associated with each of them.

At the time of this writing, FTTC and FTTH are objects of research, development, and standardization. Current deployments are in the trial stage. FTTC and FTTH, however, offer the promise of optimizing efficiency, reliability, and functionality in a way that has stimulated many telcos around the world to support these innovations.

BASIC DEFINITIONS FOR FTTx

Chapter 4, "xDSL Access Networks," noted that DSL networks—including FTTC and FTTH—can be differentiated by the point at which fiber is terminated in each case. FTTC and FTTH are on the far end of a continuum of RBB technologies in that they extend fiber the furthest—in the case of FTTH, the fiber is extended all the way to the customer's home.

Hybrid fiber coax is at the other end of the continuum. In the HFC case, fiber is terminated within two to three miles of the farthest served residence, and coaxial cable is used as the drop wire. In the case of ADSL, fiber is also terminated within CSA or RRD range, and phone wire is used as the drop wire. In the case of VDSL, fiber is terminated within a few hundred feet and phone wire is used as the drop wire.

In the case of FTTC, fiber is terminated within a few hundred feet of the residence and the drop wire is coaxial cable or phone wire. If the drop wire is coaxial cable, it is for the exclusive use of the residence and not shared with other households, similar to HFC. Additionally, the cable is not attached to the NID, but rather extends inside the home to a network terminator (NT). User applications are attached to the NT. For FTTH, fiber is terminated at an ONU that can either be inside the home, or more likely, attached to the side of the house, like the NID. No metallic wiring exists, except perhaps inside the home.

FIBER TO THE CURB (FTTC)

The term FTTC has been used to describe a variety of network architectures in which optical fiber is brought close to the residence where an *optical network unit (ONU)* couples the fiber to some form of metallic wiring that in turn leads to the residence.

The *Digital Audio Visual Council (DAVIC)* has specified four variations of FTTC, called Profiles A, B, C, and D. Profile A specifies the use of coaxial cable as the drop wire to the residence and has the highest speed. Profiles B, C, and D all permit the use of phone wire or coaxial cable as the drop wire. When configured with phone wire, Profiles B, C, and D are similar to VDSL, so they will not be discussed extensively here. The discussion of FTTC in this chapter focuses primarily on Profile A (see reference [DAVIC]).

The valid bit rates and drop wiring (wire between the ONU and the residence) for the various profiles are shown in Table 5–1.

Table 5–1 *Comparison of FTTC Profiles*

DAVIC Profile	Downstream	Upstream	Drop Wire
A	51.84 Mbps	19.44 Mbps	Coaxial cable
B	51.84 Mbps	1.62 Mbps	Coax or phone wire
C	25.92 Mbps	1.62 Mbps	Coax or phone wire
D	12.96 Mbps	1.62 Mbps	Coax or phone wire

In addition to these profiles, other variations of FTTC exist depending on the installation, packaging, and location of the ONU. There is, for example, an option to locate the ONU on the premises of a *multiple dwelling unit (MDU),* such as an apartment, cooperative, or condominium. In this case, FTTC is sometimes referred to as *Fiber to the Building (FTTB).* For FTTB, the ONU can be situated in a more environmentally friendly location than an outdoor pedestal. It also can be configured to serve more residences than the FTTC case because of the higher density of dwellings.

In another variation, the ONU is located at a roadside cabinet, or pedestal. This is sometimes referred to as *Fiber to the Cabinet (FTTCab)*. In this case, the ONU is in a more hostile environment, and probably will be servicing a fewer number of residences. Therefore, the ONU packaging for FTTCab will likely differ from FTTB.

One other nomenclature, FTTK, is not a technical variation of FTTC, but rather a linguistic one. What is a *curb* (the lateral boundary of a paved street) to Americans, is referred to as a *kerb* in the UK.

Important vendors of FTTC are Next Level Communications (specializing in Profile A), Broadband Technologies (specializing in Profile B), and Alcatel, which is investigating all options.

FTTC Architecture

Whether FTTC, FTTK, FTTB, or FTTCab, the basic elements of the technology are equivalent. Architecturally, they can be depicted as shown in Figure 5–1.

For downstream data, bits are received by the FTTC *access node (AN)* that receives ATM cells from the Internet, corporate network, or some other source. The AN serves as a function similar to the Cable Data Modem Terminal Server in the case of HFC and the DSLAM in the case of xDSL. Its job is to couple the backbone ATM network to the access network medium.

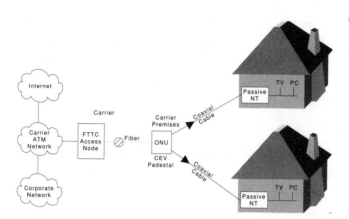

Figure 5–1
FTTC architecture.

In this case, the access medium is single-mode fiber that connects to an ONU in the neighborhood. The ONU accepts the downstream optical traffic and converts it to electronic transmission. It then transmits the signal to the residence. Unlike with HFC, the coaxial cable is not shared, except by the various users within a residence.

In the reverse path, the network terminator (NT) obtains data from TV set tops (for channel selection, for example) or from personal computers in the home and requests permission of the ONU to transmit upstream. The ONU is an active device that performs a significant amount of processing.

The function of the FTTC access node is simply to provide an interface to the core ATM network and house laser transceivers to communicate with the ONUs in the city.

Optical Network Unit (ONU)

The ONU is responsible for most of the processing in FTTC systems. It is responsible for ATM layer multiplexing, MAC protocol enforcement, and electrical to optical media conversion.

ATM multiplexing is a complex function performed in a single ONU for multiple residences. In the downstream direction, cells destined for the various residences must be demultiplexed from a common ATM cell stream and sent to the proper residence. The ONU is responsible for associating the ATM cells with the appropriate home.

In the upstream direction, a MAC protocol specified by DAVIC is used to arbitrate access from devices in the home to the ONU. After upstream traffic is received at the ONU, it is time-division or frequency-division multiplexed onto the composite link for transmission back to the CO. MAC protocol enforcement for upstream traffic for FTTC is essentially the same technique that occurs in HFC networks; but in the case of FTTC, the contention is among a smaller number of potential users, specifically, multiple users within a single residence. The ONU cooperates with digital set tops or the user applications, such as personal computers or televisions, to arbitrate congestion on the return path.

The essential physical layer function performed by any ONU is the conversion of optical signals to electrical signals.

Passive NT

The DAVIC specification for FTTC Profile A calls for the interface in the home to be a passive network terminator, or passive NT. A passive NT is responsible only for providing electrical termination of the coaxial cable and coupling it to the in-home wiring system. Passive NTs are capable of splitting, grounding, and lightning protection. They do not require electrical power and are relatively simple.

An active NT, in contrast, is capable of substantial amounts of processing to provide certain additional functions, such as packet filtering, media access control, and network management. Active

NTs are more like computers or data switches. FTTC Profile A is singular among various access networks in that it requires a passive NT. Most other access networks require an active device. The role and functions of active NTs will be discussed in Chapter 7, "In-Home Networks."

A passive NT has the advantage of being less expensive than an active NT, but the disadvantage of requiring that intrahome traffic must exit the home, go to the ONU, and return. If, for example, a user wants to take computer data and display it on the TV monitor, the bits must go through the ONU. Making this loop consumes bandwidth and gives rise to security concerns.

Principles of Operation

Well-known modulation schemes are specified for FTTC by DAVIC. The upstream path is QPSK modulated; the downstream path is QAM-16 modulated. Forward error correction is used with concatenated Reed-Solomon codewords. FTTC Profile A assumes the use of ATM cells to the residence. The application might or might not be an ATM device. If not, then additional equipment or functionality, such as a set-top box or a network interface card (NIC), would be responsible for the adaptation of the legacy protocol such as IP or MPEG into and out of ATM. The use of ATM, plus some additional overhead, bytes reduces the effective throughput in the downstream direction as summarized in Table 5–2.

Table 5–2 *Nominal Versus Effective Throughput for FTTC*

Profile	Nominal Bit Rate	Effective Bit Rate
A, B	51.84	44.544
C	25.92	22.272
D	12.96	11.136

Startup Processes

The user terminal connects to the FTTC network using a digital set top. The set top transmits a registration packet upon physical connection. The initial transmission includes a MAC address so that the ONU can associate a specific user device with a wide area address. The transmit frequency from the set top is specified in the standard so there is no exchange between the set top and the network regarding what frequency the set top should use.

After the set top registers with the ONU, the ONU assigns a 4-bit device ID to differentiate that set top from other set tops in the home. In this way, packets from the carrier can be addressed directly to a particular device in the home. One address (four ones) is used as a broadcast address. In this case, the carrier can send a broadcast message to the home and all devices will receive it. Another address (four zeroes) is used for control purposes, and no user device acts. That leaves 14 other 4-bit combinations that can be assigned, so a maximum of 14 devices in the home can use a single FTTC connection.

Media Access Control

DAVIC has specified a Media Access Control protocol for FTTC. Because FTTC is a point-to-multipoint network in the home, a Media Access Control protocol is required to arbitrate access in the drop cable portion of the return path. Given the relatively few traffic sources in the home, the short distances from them to the drop cable, and their high speeds, a simple polling mechanism is used rather than the complex schemes used for HFC. Each time it is polled and granted access, a set top can transmit one ATM cell.

FTTC Profile A Versus HFC

Obviously some similarities exist between FTTC Profile A and HFC networks and FTTC Profiles B, C, D, and VDSL. Some differences do exist, however.

The differences between FTTC Profile A and HFC are:

- Bandwidth sharing

- Applicability to broadcast television

Bandwidth Sharing

In an HFC environment, hundreds of residences share a common upstream and downstream channel. In FTTC Profile A, each residence has sole use of upstream and downstream bit rates. Specifically, each home gets exclusive use of 51.84 Mbps downstream and 19.44 Mbps upstream. If multiple TVs and PCs are in the home, they share that bandwidth for the portion of network from the home to the ONU. The TVs and PCs use time-division or frequency-division multiplexing from the ONU back to the CO.

Broadcast Television

FTTC Profile A provides 51 Mbps service to the home and 19 Mbps from the home. In the downstream direction 51 Mbps is less than two QAM-64 modulated channels. Given aggressive MPEG-2 compression, that would provide on the order of a dozen digital channels. Profile A is therefore not particularly well-suited to multichannel digital broadcast television service, such as what is available on cable or direct satellite service.

Instead, FTTC offers support for broadcasting by using switched digital video techniques. This is depicted in Figure 5–2.

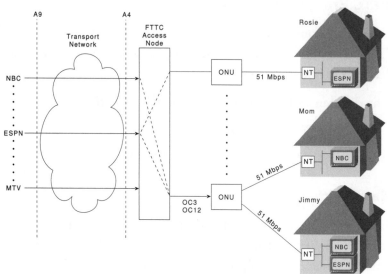

Figure 5–2
Broadcast television over
FTTC.

Multiple television channels, perhaps as many as 200, are sent to the head end/central office over ATM virtual circuits. The virtual circuits are terminated at the FTTC access node and are permanently established.

If Rosie wants to watch ESPN, she signals the set-top box in her home, which in turns sends a signal back to the FTTC access node. Channel selection is sometimes referred to as *zapping*. The access node completes an end-to-end connection between the content provider and the household. Because channel selection must give the look and feel of the type of channel changers currently used with analog TVs, the call completion process within the access node must be completed quickly.

If multiple households are watching ESPN, it is undesirable for there to be a virtual circuit between ESPN and each viewing household—this represents a waste of bandwidth. Therefore, the access node replicates ATM cells so that only a single ESPN feeds to the access node, no matter how many people in the neighborhood are watching it. In the previous figure, ESPN cells are sent to Rosie's and Jimmy's ONU.

Similarly, in a case in which multiple households share an ONU, it is undesirable for there to be virtual circuits between the access node and all households that are watching the same program. Therefore, the ONU replicates ATM cells so that only a single feed goes to the ONU from the access node for any given program, no matter how many people on the block are watching it. In Figure 5–2, Mom and Jimmy are attached to the same ONU, and both are watching NBC. The ONU replicates NBC cells for both of them. A single ONU can serve as many as 16 households.

A refinement of this architecture is network initiated channel selection. In this mode, the content provider can initiate a call to the household when it is time for viewing a particular program. The consumer selects a video on demand or pay-per-view program for a future program; and the network keeps track of the selection and initiates the VoD session at the appropriate time.

Remote tuning techniques will likely be used for broadcast TV, VoD, and nVoD applications for FTTx and VDSL access networks.

AT ISSUE

Signaling Rates at the Access Node

The challenge with switched digital video techniques is to provide the connection setup capability in the same amount of time that a subscriber would channel surf on broadcast, analog TV. To do so, the access node must be able to create and tear down connections in milliseconds. This represents an important signaling issue if the node is called upon to service hundreds of viewers concurrently. The requirement to perform signaling is likely to be the limiting factor on the side of the access node, and hence on total system economics. Here again, signaling rate is a significant barrier to system scaling.

Profiles B, C, and D Versus VDSL

Phone wire variations of FTTC bear a strong resemblance to VDSL. Both technologies employ fiber to within a few hundred feet of the home and phone wire to the home. In fact, some vendors claim that their silicon components will operate in both systems.

A few differences do exist, however. First, the standardization work for VDSL is under the control, so far, of the ADSL Forum. FTTC Profiles B, C, and D are under the control of DAVIC. Although there is a liaison between DAVIC and the ADSL Forum, the work of DAVIC is relatively settled in the areas of physical layer and link-layer protocols. The work of the ADSL Forum, however, has a long way to go. It is not guaranteed that the two technologies will track in every detail.

Second, there is discussion in the ADSL Forum about a symmetric option for VDSL that will specify bidirectional 23 Mbps service. DAVIC is not involved in any such deliberations, and is content to offer only the asymmetric versions. If a symmetric VDSL is developed, then it will likely have a shorter range than FTTC over phone wire.

Finally, VDSL will have a specification for POTS splitting, which is not specified for FTTC, and can create a great deal of difference in implementation. The passive distribution of voice will create major changes in spectrum allocation for VDSL and raise all the issues discussed in Chapter 4 regarding echo cancellation and remote powering.

Benefits and Limitations of FTTC

The main benefit of FTTC is that it offers greater speed than ADSL, HDSL, or IDSL. Because of its speed, it can support more video, primarily video on demand and near video on demand. It also has the potential for superior support of bandwidth guarantees because the DAVIC MAC protocol has a polling mechanism that avoids contention. Finally, it provides telcos with a competitive response to HFC, especially for multiple dwelling units.

On the downside, FTTC Profile A is costly because it requires coaxial cable on the drop wire, which is not in the embedded base of telcos. For Profiles B, C, and D, there is less of a speed advantage over ADSL and HDSL, and none over VDSL.

Even Profile A is insufficient in speed to support broadcast TV. Therefore, the business case for FTTC rests primarily on interactive data services.

Finally, the geographical coverage of FTTC services likely will be relatively small because the distance from the ONU to the household is so short. This will limit the introduction of FTTC services to high-density residential areas; and even there, it probably will be appropriate only for multiple dwelling units.

FIBER TO THE HOME (FTTH)

VDSL and FTTC bring fiber to within 200 meters of the home, so why not just finish the job and connect fiber all the way to the home? Fiber to the home is viewed as the holy grail of access networks. It promises high bandwidth, it is immune to electromagnetic interference, and it doesn't rust so it has lower maintenance cost for the carrier. Why not just remove the metallic elements from the access network entirely?

Until recently, there have been a number of roadblocks to extending fiber all the way to the home, including installation costs, electrical powering, a lack of applications requiring very high bandwidth, and the precision with which fiber must be handled.

Recent technical innovations and application development indicate that each of these hurdles is being confronted and the prospect of FTTH is not quite as farfetched as it once seemed.

FTTH comes in two forms. First is a shared form that looks like HFC. It has a shared forward channel and point-to-point reverse channels with a MAC protocol to arbitrate reverse traffic contention. This form of FTTH is called *passive optical networks (PON)*. The second form is a point-to-point optical service wherein each residence has a dedicated optical channel to the carrier that is shared only by the devices in the home.

Passive Optical Networks (PON)

FTTH PON is a lot like HFC. Simply replace the drop wire with single-mode fiber and reduce the number of homes per cluster from 500–2,000 to 8–32. Some of the nomenclature is different, but the startup functions of HFC are required such as ranging, the requirement for a media access control protocol to arbitrate upstream bandwidth utilization, control of transmit power from the residence, assignment of IP addresses, and packet filters.

In the case of HFC, the fiber node provided optical transmission to a neighborhood of 500–2,000 residences. In the FTTH case, a passive device called the optical splitter serves a neighborhood of 8–32 users with 155–622 Mbps of forward bandwidth and 155 Mbps of return bandwidth. The motivation for a shared media-fiber network is to introduce Fiber to the Home at the lowest possible initial cost. Specifically, the cost of the laser transmitter is shared among several residences.

Because of its lower costs, there is investigation of the use of PON by important carriers such as NTT in Japan and British Telecom (BT), both of which are conducting PON trials.

Architecture

A schematic of a PON system is shown in Figure 5–3.

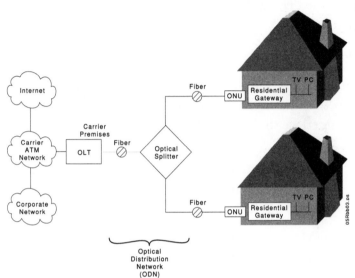

Figure 5–3
PON architecture.

For downstream data, traffic is received from the Carrier ATM Transport Network by the *Optical Line Terminator (OLT)* that receives ATM cells from the Internet, corporate network, or some other source. The OLT serves a function similar to the Cable Data Modem Terminal Server in the case of HFC, and DSLAM in the case of xDSL. The OLT's job is to couple the backbone ATM network to the access network medium.

In this case, the access medium is single-mode fiber that connects to an optical splitter. The job of the splitter is to take the downstream optical wavelength and replicate it optically so that each ONU receives the same transmission. In this way, one laser transmitter at the OLT serves multiple residences.

In the reverse path, the ONU obtains data from television set tops (for example, channel selection) or from personal computers in the home and requests permission of the OLT to transmit upstream. The splitter is a passive device and has no role in the MAC protocol.

Clearly, all this is similar to the HFC case.

Functions of the OLT are ATM switching, MAC protocol enforcement, physical layer functions, access network security, and interface to the ATM Transport Network.

- **ATM switching.** A single OLT will be connected to multiple optical splitters. The OLT is required to take traffic to and from the splitters and to route it to the proper ATM port and ATM virtual channel.

- **MAC protocol enforcement.** The OLT and the ONU must cooperate to arbitrate congestion on the return path. They do so by enforcing a MAC protocol.

- **Physical layer functions.** The OLT is responsible for optical coupling to the Transport Network. It obeys certain conventions regarding optical interfaces, such as laser launch power, frequency, and so on.

- **Access network security.** Most of the intelligence required by the network will be in the OLT. This includes packet filtering and the enforcement of connection authentication, that is, refusing to accept packets from unauthorized senders or receivers.

Optical Distribution Network (ODN)

Within the larger PON architecture, the optical distribution network is comprised of the fiber distribution plant and the passive optical splitter. The components of the optical distribution network are labeled in Figure 5–3. It provides the optical transmission medium for the physical connection of the ONUs to the OLT. The ODN offers one or more optical paths between one OLT and one or more ONUs.

Certain physical characteristics of the ODN are being standardized in the *International Telecommunications Union (ITU)*.

The transmission medium consists of one or two single-mode fibers in accordance with ITU-T Recommendation G.652. Bidirectional transmission is accomplished by use of either a wavelength division multiplexing (WDM) technique on a single fiber, or two unidirectional fibers.

In the one fiber system, the forward path and return path share the fiber by using different frequencies. The downstream wavelength is in the range of 1530–1570 nm. The upstream wavelength is in the range of 1280–1340 nm.

Optics, such as electronics, suffer from attenuation. In the case of PON, the specification is that path loss is less than 10–25 dB (per ITU-T G.982 Class B). Furthermore, the ITU specifies that the variation in path loss from one home to another be bounded. That is, one home may be closer to the OLT than another home connected to the same OLT; the standard limits the difference in path loss between the two homes to 15 dB.

The distance between the OLT and ONU can be as long as 20 kilometers.

The speeds specified by ITU are summarized in Table 5–3. Keep in mind that these speeds are shared among the homes connected to an OLT.

Table 5–3 *ITU Speed Specifications for PON*

	Upstream	Downstream
Standard	155 Mbps (OC-3)	155 Mbps (OC-3)
Aggressive	155 Mbps (OC-3)	622 Mbps (OC-12)
Lowest Speed	25.92 Mbps (ATM 25)	155 Mbps (OC-3)

Optical Network Unit (ONU)/Residential Gateway

The ONU provides the necessary functionality to connect the carrier fiber to the media in the residence. The ONU is assumed to be on the outside of the residence, similar to the placement of the NID, on the side of the house. The ONU coupled with a more intelligent Residential Gateway (refer to Figure 5–3) will generally perform the functions shown in the following list.

- **ATM multiplexing.** Multiple devices in the home will be connected to a single ATM port from the home to the carrier. The ONU and Residential Gateway are responsible for multiplexing the sessions on the fiber link.

- **MAC protocol enforcement.** The OLT and the ONU must cooperate to arbitrate congestion on the return path. They do so by enforcing a MAC protocol.

- **Physical layer functions.** The ONU is responsible for electrical to optical media conversion in the home. The home, for example, may use phone wire, coaxial cable, or even a wireless connection to connect to televisions and personal computers. The ONU would be responsible for the media translation.

- **Interface to the home network.** Home networks such as IEEE 1394 and IEEE 802.11 require some processing of their own, including protocol translation.

Because of the complicated nature of the interface functions, a component called the *Residential Gateway* or *Home Network Gateway* will decouple the purely optical functions of an ONU from high-layer protocol functions. The concept of the Residential Gateway will be discussed in more detail in Chapter 7.

Principles of Operation

All current PON proposals assume the use of ATM cells to the residence. The application may or may not be an ATM device. If not, then the Residential Gateway would be responsible for the adaptation of the legacy protocol such as IP or MPEG into and out of ATM.

The downstream signal is broadcast to all ONUs on the PON. Each upstream transmission from each ONU is controlled by OLT. A single residence can receive a variety of services, such as data retrieval, television reception, and telephony. Each of these cells is transmitted using TDM techniques. The ONU demultiplexes the optical line format into the various services.

In the return path, credits are granted to each subscriber, which permit the use of upstream time slots. Either polling or contention-based Media Access Control protocols can be used, similar to those used by FTTC and HFC. The use of time slots mean that all home transmitters connected to a common OLT use a common upstream frequency. Therefore, a single-laser receiver at the central office can receive traffic from all residences. This is an important cost-saver and is why shared FTTH is seen as more practical than point-to-point FTTH.

On the other hand, if the cost of laser components declines, it is possible to consider the case of dedicated fiber connections to the home.

Broadcast television will be supported using switched digital video techniques, just as in the FTTC case. Even though the return path speed of FTTH is faster, the speed at which tuning occurs is still an issue because the critical factor is how fast the access node can create the virtual channel from the content provider to the subscriber.

Point-to-Point FTTH

PON is a shared topology. Up to 32 residences can share 622 Mbps of downstream bandwidth and 155 Mbps of upstream bandwidth. If all residences are active, average bandwidth per residence will be far less than HFC. This would seem to be disappointing given the allure and reputation of fiber to the home.

As more and more residences subscribe to FTTH, the cluster size will drop from 32 down to as few as 8 because service providers will have proper funding to penetrate the splitters farther into the neighborhood. Ultimately, each home may have a dedicated fiber circuit to the carrier.

Many companies are studying ways to use *Dense Wavelength Division Multiplexing (DWDM)* to create private virtual paths into the home. DWDM-based systems will enable carriers to allocate wavelengths to customers, providing dedicated bandwidth of up to 155 Mbps in both directions to each home. Figure 5–4 illustrates the use of DWDM in a point-to-point FTTH environment.

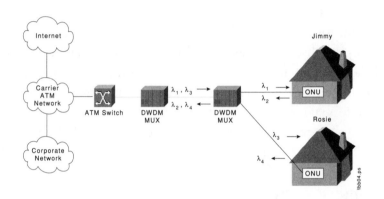

Figure 5–4
Point-to-point FTTH architecture.

As shown in Figure 5–4, content from the Carrier ATM Network is received at an ATM switching function. In the forward direction, content going to Jimmy is sent on one frequency, and content going to Rosie is sent on another. These frequencies are multiplexed onto a single-fiber loop by a DWDM. At the other end of the loop, the frequencies are demultiplexed and forwarded to the respective households.

A DWDM is basically a frequency division multiplexer for optics. Different wavelengths, or frequencies, travel simultaneously between the multiplexers. The subscriber will be allocated an upstream and downstream frequency for his or her exclusive use. In the scenario illustrated in Figure 5–4, Jimmy receives on frequency λ_1 and transmits on frequency λ_2; Rosie receives on frequency λ_3 and transmits on λ_4.

Current versions of DWDMs multiplex 16 frequencies per fiber. Certainly, this number will increase quickly. But for now, DWDM is used exclusively in carrier backbones because of cost of the lasers. Currently, each frequency requires a separate laser. To have a point-to-point FTTH solution, there needs to be greater cost reductions of multi-frequency lasers.

With point-to-point, there needs to be a laser transceiver at the head end for each subscriber. Furthermore, because of possible changes in subscriber usage, it will be necessary to remotely control the transmit and receive frequencies of the laser transmitter in the home. Developments required to reduce laser costs sufficiently to make point-to-point FTTH competitive with HFC or PON are expected to be implementable sometime between the years 2000 and 2002.

Benefits of FTTH

FTTH is viewed as the end game for access networks for a variety of reasons.

Speed

Residences will benefit from more bandwidth, especially in the case of DWDM. In the case of both PON and point to point, FTTH offers the greatest amount of upstream bandwidth compared with HFC, xDSL, or FTTC, and thus it is the most appropriate technology for residences to host their own web sites. Businesses can take advantage of high-speed return for corporate web hosting.

Remote Provisioning

The key benefit of FTTH for the carrier is its capacity for *remote provisioning*. Remote provisioning refers to the capability of the carrier to configure customer service from the CO. A consumer, for example, can elect a service level agreement that includes 50 Mbps service. Later, the same customer might want 100 Mbps service and later still, as little as 2 Mbps. Because the bandwidth to the consumer is greater than any of these options, the carrier need only impose some flow control measures at the CO to restrict consumer bit rate to the requested service level. Eventually, given improved network management techniques, consumers could be able to conduct provisioning for themselves.

The carrier or the consumer can configure bandwidth for exclusive use for relatively long periods of time (hours, or even days). Remote provisioning would be subject to available link capacity and pricing, but many of the operations and maintenance costs could be reduced.

Remote provisioning of bit rate is enabled by the use of ATM to the home. ATM has the capability to support usage requests expressed in different semantics. Such semantics include peak rate bandwidth, average bandwidth per unit time, and the variability of transmission delay. Of course there would be monetary costs associated with such flexibility, but this would offer greater control for the user and possibly less provisioning cost to the carrier.

Carrier Cost Reductions

The key benefit of FTTH for the carrier is reduction in costs. These reductions come from powering, self provisioning, and the reduced costs of outside electronics and cabling. The cost of power to the ONU will be passed to the customer because it is on the consumer premises. A later section, "Power to the ONU," discusses this in more detail.

Furthermore, it is a relatively smooth transition from ADSL to FTTC to FTTH. The carrier can start with lower-speed technologies and proceed to higher speeds by pushing the ONU farther into the neighborhood. This can be a measured process that can be funded in part by the maintenance budget of the LECs, which enjoys regulatory relief.

Greater Reliability

Optical components are immune to RF interference, such as ingress and crosstalk. They also don't rust, which extends the usable life of components in the field. Optical signals take longer to attenuate than electrical signals through metal, and there is less corrosion. This all suggests that optical signal quality is superior to that of wired infrastructures, and will likely remain so as the infrastructure ages. Fiber also has fewer active repeaters than metalic networks for equivalent distance. This contributes to greater reliability.

Full Service Networking

FTTH and HFC are alone in their capability to provide *full service networking (FSN)*. FSN refers to the capability to provide high bandwidth, multichannel services, such as broadcast television, and narrowband services, such as telephony. Marketers dub this feature *futureproofing*, which implies the capability to accommodate whatever services and content the future holds without changing the infrastructure. Over time, therefore, life cycle costs are minimized.

Challenges to FTTH

Although the prospect of sharing 622 Mbps among a handful of homes is enticing, the fact remains that PON offers less bandwidth in the forward direction than HFC for broadcasting. Furthermore, HFC is operational in millions of homes, whereas FTTH is only in the trial stage. Therefore, FTTH, in either PON or point-to-point form, will not have the breadth of programming that DBS and HFC have. It is likely that FTTH will have an advantage for pull-mode services, either VoD or pull-mode data, but it remains to be seen if push-mode or pull-mode services will provide the most lucrative funding model for RBB.

There are other challenges to FTTH as well.

Fiber Handling Problems

Chapter 2, "Technical Foundations of Residential Broadband," overviewed some technical issues regarding fiber. In the context of FTTH, handling and bending of fiber are of particular concern. Fiber is a difficult medium to handle. When joining two segments of fiber-optic cable end-to-end, the part that actually transmits light is only 50–62.5 microns in diameter (a micron is one millionth of a meter). Like railroad tracks, if the fiber segments are not perfectly aligned, severe impairment results. Splicing fiber cable requires expensive precision equipment. Technical innovations in development will increase the tolerances with which fiber can be handled, but these need further refinement. Bending is also a consideration—fiber cable cannot be bent excessively without incurring severe attenuation.

In general, electrons traveling through metal wire are much more forgiving than photons traveling through fiber. The increased precision and costs associated with fiber installation as compared to metal wire are ongoing challenges to FTTH.

Startup Costs

As noted earlier, FTTH offers operational cost savings to carriers, including the reduction of their electric bills, reduced provisioning costs, and reduced costs of outside electronics and cabling.

These operational costs, however, are long-term benefits, which are realized during the life cycle of the service.

Major costs are associated with the startup of service. Among the key costs are installing the fiber (which involves digging), connecting the subscriber equipment in the home (which involves labor and precision equipment), and the subscriber equipment itself. Central office equipment costs have decreased because reductions in the costs of lasers and wavelength division multiplexers have been considerable in recent years. But, more cost reduction is required to make subscriber equipment FTTH costs competitive with wired infrastructures. This is especially true for the laser transmitters in the home, which basically become consumer devices.

For the moment, the startup costs outweigh the operational cost savings and might continue to do so for years to come.

Power to the ONU

Another requirement for widespread FTTH is accord among regulatory agencies, telcos, and consumers regarding electrical power for the ONU. The ONU is on the residential premises and must derive its electrical power from the consumer or possibly from the network. But network powering is cost-prohibitive for the carrier, because it would mean the installation of metallic wire to the home just for the supply of electrical power. This wiring would be in addition to network wiring. For FTTH systems to be economically viable, the ONU needs to be powered from the consumer's outlets, with a battery backup provided either at the consumer's or at carrier's expense. Whether or not this can pass muster with state regulatory agencies or the public at large is to be determined.

There will be arguments, from the carriers primarily, that the residential powering problem is diminishing. The use of cellular phones is considerable, and they can be used in an emergency. Other alternatives are batteries, which can operate phones for 8 to 24 hours.

Encryption and Wiretapping

Because the cable is shared, data likely will be encrypted. Encryption, and the fact that fiber is more difficult to wiretap than metallic cables, means that some method of enforcing court-ordered wiretaps must be implemented before widespread use of FTTH will be allowed in the United States.

Already, conflict has begun between law enforcement and the telecommunications industry over wiretapping of digital telephony. Techniques used to wiretap analog voice calls do not work when voice becomes a digital application. Law enforcement is asking that modifications be made to digital telephone equipment to facilitate wiretapping. These modifications involve some costs, resulting in objections from telecommunications carriers. This issue is yet to be resolved. It is a harbinger of things to come, when all transmission to and from the home is digital.

VIEWPOINT

Regulatory Issues for FTTH

Powering to the ONU is one issue the carriers might eventually win. The venue for Universal Service might change from wired to wireless services over time. Offering analog cellular or digital personal communications service as an alternative to wireline Universal Service might be in the public's best interest as well as being cheaper for the FTTH carrier. If such a transition is approved by regulatory bodies, then network powering will no longer be required.

continues

On the other hand, the argument over encryption and wiretapping is one that the carriers will not win. There is widespread public support for law enforcement. If law enforcement can demonstrate that criminal activity uses digital facilities, regulation and legislative support would follow. Encryption and wiretapping increase costs, especially on the subscriber equipment, and a number of engineering issues are to be determined, such as key exchange, encryption algorithm, and law enforcement access to CO equipment.

SUMMARY

As candidates for residential broadband networks, FTTC and FTTH seem to represent the best of all possible worlds. They hold out the promise of being full-service networks—the elusive single entity that can provide all the high speed and traditional services that consumers are likely to want delivered to their homes. These technologies are also the least established and applied at this point in history, so predicting their future is difficult. Table 5–4 summarizes the challenges to facilitators of Fiber to the Home.

This chapter concludes with the consideration of specific wired networks. The next chapter turns to wireless networks, including one of the most well-established residential broadband candidates so far, direct broadcast satellite.

Table 5–4   *Central Issues for FTTx and RBB*

Issue	Challenges	Facilitators
Media	Increased fiber distribution entails digging; requires precision handling, equipment, and training.	Fiber is being installed anyway and will reduce life cycle maintenance costs for carriers.
Modulation schemes	QAM 16 and QPSK need to operate at high speeds.	Digital modulation technique for fiber is a simple on/off mechanism.

Table 5–4 *Central Issues for FTTx and RBB, Continued*

Issue	Challenges	Facilitators
Noise mitigation	Higher speeds create little tolerance for line impairments.	Fiber is more immune to external noise.
Signaling	Switched video signaling must mimic home channel surfing.	FTTx sessions don't connect through a voice switch, thus reducing signaling through the voice switch.
Data handling capabilities	Speed is less than cable in the forward path for all but point-to-point FTTH.	Good return path service enables reliable interactive service with known QoS.
Standards	Lagging behind more established technologies. In particular, standards for end-to-end service architectures are unresolved.	Substantial agreement on important PHY level issues exists. Relatively few bodies are deliberating FTTx specifications.
Business/ Financial	Infrastructure costs; telcos will need to install either coaxial cable or fiber. Telcos do not have a good track record at rolling out high-speed services. How to transition current customers' rewiring. How to provide existing digital services, such as X.25 and frame relay.	Reduces telcos' real estate need. Reduces data burden on voice switches and need for new one. Maintenance budget can be used for certain upgrades. There is a logical transition plan from ADSL through FTTC to FTTH. Reduced carrier maintenance because powering is the customer's responsibility (FTTH).

Table 5–4   *Central Issues for FTTx and RBB, Continued*

Issue	Challenges	Facilitators
Market	Questionable support for broadcast video.	If switched from full-service network capabilities, digital video will perform. Enough bandwidth to support customers' web sites.
Regulation	Mechanisms to support law enforcement wiretapping not yet resolved (FTTH). Uncertain if regulation will force carriers to provide network powering to the ONU; this will increase costs (FTTH).	Possible relief from powering requirements if Universal Service can be provided via wireless telephony.

REFERENCES

[DAVIC] Digital Audio Visual Council, "DAVIC 1.0 Specification, Part 8, Lower Layer Protocols and Physical Interfaces." September 1995. Also at http://www.davic.org.

Effenberger, Lu. "Overview of FTTH Networks, Past History, Current Status, Future Designs." Broadband Access Systems, Wai Sum Lai, Sam Jewell, Curtis Stiller Jr., Indra Widjaja, Dennis Karvelas, Editors. *Proceedings, SPIE 2917.* 1996, pages 293–304.

Faulkner, David, et al. "Full Services Access Networks Initiative." *IEEE Communications*, April 1997.

Gibbons, Mark, Nguyen, J.J. Werner, Tom O'Shea, Ken Buckland, "Clarifying the Definition of FTTC." *ATM_Forum/95-1037*, August 1995.

Takigawa, Yoshihiro, NTT Optical Labs, Alan Quayle, BT Labs. "A Universal ATM Passive Optical Network Specification." *ATMF Contribution 97-242*, May 1997.

Townsend, Rick, J. J. Werner, Gang Huang, Mai-Huong Nguyen (AT&T Bell Laboratories). "Channel Impairments for Local Distribution Systems of FTTC Networks." *ATM_Forum/95-0614*, June 1995.

Vos, Eric. Alcatel Telecom, "Proposal for an ATM-PON Interface Specification." *ATM Forum 97-007*, February 1997.

CHAPTER 6

Wireless Access Networks

The previous chapters point to the tremendous strides recently made on wired networks. Copper and fiber networks are becoming more ubiquitous, faster, easier to use, and less costly. Furthermore, the transmission capacity of wired networks is considered infinite, because carriers arbitrarily can manufacture bandwidth as demand increases.

Despite the capacity of wired networks, wireless networks have had the greatest success among consumers. Broadcast television, cellular telephone, paging, and direct broadcast satellite are all wireless services that have met with commercial success—even despite the fact that wireless networks typically carry lower bit rates and higher costs than wired networking. More than 39 million consumers, for example, subscribed to cellular telephones as of the end of 1996—in spite of variable voice quality and high costs—to have the ease of use of a wireless network with no cabling and the freedom to be mobile while still maintaining access to the service.

With cabling, you could dig in the wrong place, such as in places where relatively few takers exist. Product managers who roll out wired services struggle with marketing and demographic studies to determine the best neighborhoods in which to introduce services. In the case of one cable company, for example, the choice of head ends to initiate a data/cable service was determined in part by identifying Prodigy online service users.

Even if the right neighborhoods are identified, it is expensive and time-consuming to dig or install overhead cables. Furthermore, permits and easements must be obtained. To some observers, the fixed networks of wired systems look like vulnerable high-capital assets in a world of fast-changing technologies.

A number of wireless Access Network technologies are intended by their proponents to serve the consumer market. These are *direct broadcast satellite (DBS)*, *Multichannel Multipoint Distribution Services (MMDS)*, *Local Multipoint Distribution Services (LMDS)*, and *Low Earth Orbit (LEO)* satellites. Each will be considered in this chapter.

REFERENCE ARCHITECTURE FOR WIRELESS NETWORKS

Figure 6–1 illustrates how wireless networks map to the DAVIC reference model discussed in Chapter 2, "Technical Foundations of Residential Broadband." The return-path flows, if any, travel through wired networks or, in the case of LMDS, through wireless networks.

Figure 6–1
Wireless reference
architecture.

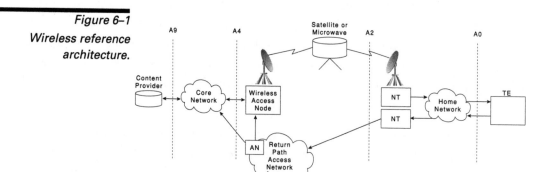

The content provider forwards content through the Core Network and to the wireless access node. This access node reformats data and modulates it for satellite or land-based microwave transmission. A receiving antenna at the home end forwards traffic through the Home Network to the terminal equipment, which is either a TV set-top box or a PC.

In the return path, the consumer uses either the same network used for the forward transmission or another Access Network. Another Access Network is needed when using DBS or MMDS services, which are one-way networks. The return-path network could be telephone return, xDSL, or another wireless service, such as digital *personal communications services (PCS)*. PCS is a digital service to be offered on recently auctioned spectrum. It is expected that PCS service will include wireless voice—a digital form of cellular telephony—as well as wireless data. But PCS also can be adapted to carry wireless data service.

Because forward- and return-path traffic can use different physical media, traffic sources must be matched so that a single bidirectional session exists between the content provider and the terminal equipment.

This matching can be performed by the wireless access node or another switching/routing device inside the Core Network.

THE MOTIVATION FOR WIRELESS

Each nation's airwaves historically have been a tightly regulated commodity for a variety of reasons. Spectrum often was underutilized because older modulation techniques used bandwidth inefficiently, unlike today's digital modulation schemes. In the United States, the FCC granted spectrum licenses for free until 1993, so there was little incentive to develop new modulation technology. Moreover, a good portion of the available spectrum is reserved for either high-priority uses by local agencies such as fire and police departments or national uses such as Federal Aviation Administration air traffic control, deep satellite telemetry, and military communications. Broadcast TV and radio also consume choice swaths of spectrum and further limit the airwaves, and finally, very high frequency spectrum (above 10 GHz) was deemed unusable until recently.

Enter technology. With the advent of digital technologies, spectrum suddenly has become more plentiful, and even abundant. First, spectrum is being used more efficiently, primarily because of the development of digital modulation and compression techniques. Second, very high frequencies that were previously unusable now are becoming usable due to improved transmission techniques. Added to these technological advances, policy changes that relocate incumbent license users to less valued spectrum force both the new bandwidth services and present license holders to use spectrum more wisely. The recent auctioning of spectrum by the FCC puts a price on bandwidth and thereby creates incentive for more efficient use. All these forces combine to make wireless the technology of choice for new entrants to broadband. These new entrants include existing service providers who want to enter someone else's market.

Service providers of all types have become involved because of the numerous advantages of wireless networks. These networks involve the lowest startup cost and therefore the lowest risk for new services experiencing a low take rate. Wireless networks also have the fastest rollout in that such networks can pass more residences in a shorter period of time than wired infrastructures. Furthermore, wireless has the lowest maintenance cost—air doesn't rust or need replacement.

On the other hand, wireless networks use even higher frequencies than a wired infrastructure. The result is that wireless has not been reliable; in fact, doubts about its robustness continue. But the success of other wireless consumer products and the difficulties with wired infrastructure have resulted in continued product development in a variety of wireless broadband technologies.

DIRECT BROADCAST SATELLITE (DBS)

Although cable operators were only talking about digital TV, *direct broadcast satellite (DBS)* companies actually achieved it, taking the cable industry by surprise. Early entrants were Primestar, DirecTV, and United States Satellite Broadcasting (USSB), all which were launched in 1994. Echostar launched in March 1996, and recently Rupert Murdoch's News Corporation—which has launched DBS services BSkyB and JSkyB (for Britain and Japan Sky Broadcasting, respectively)—announced the formation of *ASkyB (American Sky Broadcasting)* in partnership with MCI. To acquire the orbital slot its satellite will occupy, MCI paid $682 million; the company will pay another $150 million for each satellite, as well as $50 million or so to ensure the rocket launch, and hundreds of millions more for content and marketing. In a recent sign of the times, News Corp. signed a deal with Primestar in which News Corp. offers its orbital slot and two satellites to Primestar in exchange for a $1.1 billion equity stake and the guarantee that Primestar will carry Fox network content.

In the United States, DBS is viewed as a commercial success and has signed a surprising five million customers in its first three years of operation. This signup is particularly strong considering that customers initially paid up to $800 for a home satellite dish and installation. Such a strong start has cable TV operators concerned. Even more troubling for U.S. cable operators is that the average DBS subscriber spends about 50 percent more per month than the average cable subscriber (about $52 versus $35 per month). This is due to the upscale sale of DBS and the sales of premium sports and movie packages.

In fact, much of the success of DBS is due to imaginative programming packages. In particular, aggressive marketing of sports packages (including college basketball and professional football) has created differentiated content for which DBS has found a ready market.

In addition to U.S. businesses, foreign participation in DBS is substantial, particularly from DirecTV and News Corp. Table 6–1 lists various DBS services worldwide. The orbital positions of the Canadian and Latin American efforts can serve the United States as well.

Table 6–1 *Worldwide DBS Networks*

Participant (Country)	Channel Capacity	Description
Shaw (Canada)	120	A Canadian cable operator trying satellite.
Galaxy (Latin America)	144	A consortium with DirecTV.
Televisa (Mexico)	100	A network of News Corp., TCI, and PanAmSat.
BSkyB (UK)	140	Rupert Murdoch in the UK.
British Interactive Broadcasting (UK)	140	An interactive version of BSkyB in partnership with British Telecom, Midland Bank, and Matsushita.
Televisa (Spain)	140	An extension of Latin American service of PanAmSat to Europe.
DF1 (Germany)	50	A network that is obtaining some funding from BSkyB.

Table 6–1 *Worldwide DBS Networks, Continued*

Participant (Country)	Channel Capacity	Description
TPS (France)	100	A network of France, Telecom, and others.
JSkyB (Japan)	100	Rupert Murdoch again, this time with Softbank and TV Asahi, a major Japanese broadcaster.
Star TV (All Asia)	100	Rupert Murdoch again, this time by himself. Now an analog service migrating to digital. Problem: How to replace all those set tops in India?
PerfecTV (Japan)	58	A consortium of Japanese trading companies.
DirecTV (Japan)	100	An extension of U.S. service to go head to head against Rupert Murdoch.
Agila (Philippines)	160	A consortium of PTT (the national phone company) and other national firms with Aerospatiale to launch three satellites by 1998.

In Europe, DBS has relatively few subscribers so far, but major companies are planning market entry. Asia and the Middle East also have become more aggressive in rolling out satellites. A major reason is that a lot of islands preclude cabling, and the populations are so enormous that time to market is essential. It is too hard to cable the Philippines, for example; satellite is the only way to go.

DBS Architecture

Architecturally, DBS is a simple concept. As shown in Figure 6–2, DBS operators receive analog TV reception from the various networks at a single giant head end. DirecTV's head end, for example, is in Castle Rock, Colorado. The analog programming is encoded into MPEG for digital retransmission. A control function regulates the amount of bandwidth accorded to each MPEG stream and determines how the MPEG knobs (control parameters)—such as group of pictures length—are specified.

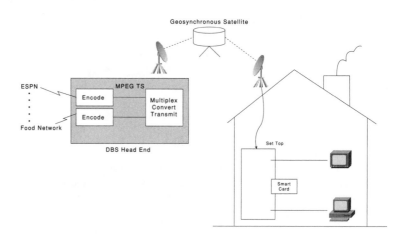

The settings of the knobs are closely guarded secrets of the DBS operators. ESPN, for example, tends to require more bandwidth than the Food Channel. ESPN has a lot more motion and, more importantly, greater viewership and advertising revenue. How much more would ESPN pay for access than the Food Channel? How much extra bandwidth is ESPN getting, and for how much? What MPEG knobs should the carrier use, and what knobs does its competition use? This is not public information.

ESPN, the Food Channel, and all other channels are encoded into MPEG transport streams, multiplexed together, and then converted to the uplink frequency.

The composite signal of 140–200 channels is transmitted up to a geosynchronous satellite. The major North American geosynchronous satellites for DBS so far are placed at longitudes 85 degrees west (Primestar), 101 degrees west (DirecTV), and 119 degrees west (Echostar). The Primestar slot rests on the longitude that passes through the East Coast of the United States, the DirecTV longitude bisects the center of North America, and the Echostar longitude passes through the West Coast. From these orbits, each satellite can broadcast over the contiguous United States, southern Canada, and Mexico.

The satellite receives transmission and remodulates to the designated spectrum for DBS. DBS occupies 500 MHz in the 12.2–12.7 GHz range, within what is known as the *Ku Band*. DBS satellites are allowed by regulation to broadcast at a higher power (120W) than the larger C-band satellite dishes currently used by many households to enable reception on small satellite dishes. This higher powered transmission and smaller dish distinguish DBS from other forms of satellite reception.

The DBS uses QPSK modulation to encode digital data on the RF carriers. DirecTV encodes using MPEG-2 syntax to enable a density of up to 720×480 pixels on the user's monitor. Primestar uses a proprietary video compression system developed by General Instruments called DigiCipher-1. Echostar uses a transmission system based on the European *Digital Video Broadcast (DVB)* standard. DVB uses the standard MPEG-2 and standardizes control elements of the total system, such as conditional access.

Although 720×480 is the maximum resolution offered today, DBS is capable of higher pixel resolution. In fact, DBS could be an early delivery vehicle for HDTV programming. Most analysts think DBS will prove to be the most cost-effective means of delivering HDTV to homes in the United States for years to come. This is because most HDTV content is expected to be films—which are easily converted to HD and which the DBS vendors routinely offer—and the cost of upgrading to digital for DBS is limited to the head end and the receiving dishes. For over-the-air broadcast-ers, upgrading to digital means that each individual broadcast affiliate must make an investment.

On some systems, a credit card–sized plug-in board called a *smart card* is used for key management of conditional access sys-tems. The smart card plugs into the decoder, and without this card there is no reception. When the conditional access system is compromised by hackers, the carrier can mail replacement smart cards with new encryption keys to subscribers. This costs around $10 per card but is a little more secure than other systems.

AT ISSUE

Security Using Smart Cards

Debate continues about the security and expense of smart cards, but the cards are slowly gaining momentum despite the cost of smart-card read-ers in set tops. Because carriers are reluctant to replace entire set tops, they typically live with breaches of conditional access. Smart cards give carriers an option, though, and raise the bar slightly against hackers. Of course, hackers are an ingenious lot and no doubt eventually will crack smart cards as well. The question is whether the cost of maintaining con-ditional access is worth the effort.

European carriers think so. Smart cards will be used in Europe to help DBS service providers overcome a problem not found in the United States. Because of the proximity of national boundaries, and because the issuance of program licenses is sometimes limited by country, it is nec-essary to provide conditional access by country. A DBS provider, for ex-ample, might have the right to transmit a certain program in the United Kingdom but not in Spain, where another DBS provider has the program license. In this case, a U.K. resident on vacation in Spain should not be able to use his U.K. smart card there. Measures are under consideration to provide this extra security. This can be done, for example, by combin-ing information on the smart card with information about the set top into which it is inserted.

Another use of smart cards is for electronic commerce. Credit informa-tion could be infused onto a smart card so that commercial transactions can be made online without the need to reveal credit card information. If this form of commerce expands, the costs associated with smart cards will fall, making it more attractive for purely conditional access use.

Most, if not all, of the decoders contain a high-speed data port that can be connected to a computer or another external decoder. One such service is *AgCast*, offered by Echostar. AgCast is a one-way service that provides agricultural information such as weather and crop reports to rural subscribers. Data appears rather like that generated by a stock ticker service.

The set-top hardware used by DirecTV is called *Digital Satellite System*, or *DSS*™. Sony, Thomson Consumer Electronics, Hughes Network Systems, Toshiba, Matsushita (Panasonic), and Uniden all sell the DSS receiving equipment.

Challenges to DBS

DBS sales started out impressively but have slowed, and some experts believe that DBS penetration will not exceed 10 million to 20 million homes in the United States. DBS still loses money and faces keen competition from a strengthening cable industry and other wireless technologies, which are discussed later in this chapter. Also on the horizon is digital over-the-air broadcasting. Alphastar, the smallest DBS operator, declared bankruptcy in June 1997. Consolidation among service providers certainly will take place, leaving at most three and, more likely, two national providers in the United States.

Data Service

DirecTV has partnered with Microsoft to produce a push-mode data service over DBS. Supporters refer to the service as "Point-Cast on steroids." The service broadcasts roughly 200 popular web sites, which are cached in the consumer's PC. Some content will be cached at the service provider's site. Instead of having a point-to-point connection with the Internet, consumers access content on the hard drive or service provider cache. In addition to web sites, other data services such as AgCast or stock quotes can be offered, either by continuous feeds or by caching on the consumer's PC. The problem with this model is that you cannot access a web site that is not part of the service, because no point-to-point return-path connection exists.

One form of point-to-point data service, called *DirectPC*, can reach the Internet. DirectPC is jointly owned by DirecTV and Hughes Network Systems. DirectPC reserves 12 Mbps of downstream service and uses a telephone as a return path.

Because the *footprint* (the portion of the Earth's surface covered by the signal from a communications satellite) is so large for geosynchronous satellites, it is possible that thousands of users will want to use the common 12 Mbps of service concurrently. The more concurrent users there are, the less bandwidth each user gets. To provide a balance between bit rate and the number of concurrent users, DirectPC offers roughly 400 KBps of service to 300 concurrent users. Although DirectPC is a high bit rate service, it can't scale to many thousands of users because of the national footprint.

To serve more users, a mechanism is required to reduce the satellite footprint. A technique called spot beaming is designed for this purpose and will be discussed in the following section, "Local Television Content." Think of spot beaming as functioning as clustering does for cable systems—it's basically a way to segment users into smaller groups.

Both data services offered by DBS are limited in their capability to compete with other Access Networks for high-speed Internet access because of the one-way nature of the medium. DBS could have a role for push-mode data, which is what the Microsoft/DirecTV product is targeting.

Local Television Content

Local TV programming is important to viewers. Morning ratings are high because viewers want local weather and traffic to start their day and get to work. Local sports are also important in many communities; high school football is popular in Texas, for instance, while high school basketball is important in Indiana. Another local content issue arises due to time zones. Many people on the West Coast, for instance, are unhappy that some DBS programs are received three hours early because they originate from the East Coast. Finally, local advertisers require stations to broadcast only locally—it does a local restaurant no good to advertise nationally.

The delivery of local content on DBS is hampered by two issues: one technical and one legal. The technical issue concerns how to reduce the footprint from a geosynchronous satellite. This problem is being addressed by the development of a frequency reuse technology, called *spot beaming*. With spot beaming, the satellite can broadcast to multiple small footprints rather than a single footprint covering the entire country. This is done using sophisticated antenna-shaping techniques originally developed for military satellite applications.

Spot beams from satellites can segment the country into roughly 32 geographic areas per satellite. Plans call for spot beaming satellites to transmit from eight transponders with four spot beams per transponder at a cost of $250 million per satellite.

With 32 geographical areas, spot beaming provides less granularity than current local broadcast. With 20–40 footprints, the population of the United States is divided into regions with roughly 6 million to 12 million people per footprint. It remains to be seen if this is sufficient granularity to keep advertisers and viewers interested. The technical problem with spot beaming is that new satellites are required. At hundreds of millions of dollars per satellite, including launch fees and insurance, this is a major commitment.

Perhaps a greater problem is the treatment of royalties. When a local station broadcasts, its content is copyrighted. With spot beaming, a local broadcaster's range is extended geographically and sometimes exported to other areas. It could, for example, be possible to broadcast stations in Los Angeles county to neighboring counties, which already have separate affiliates for the same network.

In this case, the local broadcaster is paid a copyright fee. The *Satellite Home Viewer Act (SHVA)* provides a compulsory right for satellite providers to retransmit network signals, and the act provides a royalty fee for that right. Satellite providers pay six cents per subscriber per month to the programmers and local stations for that right. Programmers feel they should receive more like 35 cents, though, and as of this writing, negotiations are in progress to form a new royalty agreement. Without an agreement, it is unclear which programs will be retransmitted and for how much.

Other provisions of SHVA preclude the retransmission of local signals into areas in which the signal can be received over-the-air. A CBS affiliate in one area certainly would not want a satellite provider spot beaming another CBS affiliate into the same geographic area. Spot beaming does not have the granularity to circumvent the current SHVA prohibition on retransmission of local signals. Therefore, DBS vendors are lobbying Congress to amend the provision. These negotiations regarding royalties and copyrights are expected to be a part of a renewed SHVA after the current SHVA expires in 1999.

Another aspect of local content that was the object of legislative discussion was the applicability of Must Carry rules to DBS vendors in the event that they use spot beaming. Without spot beaming, requiring DBS vendors to honor Must Carry rules would be technically infeasible, because it would mean carrying 1,600 local channels on a DBS service. But with spot beaming, DBS service looks very much like cable, so should Must Carry apply? The thrust of current discussion is that the rule should apply. In fact, News Corp. has stated before Congressional hearings that it would offer no objection to this principle. Details remain uncertain, but it appears that the major legislative and regulatory issue for DBS is SHVA, not Must Carry.

Multiple Home Viewers

TV customers currently have cable-ready VCRs and TVs; they can record on one channel and view another simultaneously. DBS systems can decode only one channel at a time, so a separate decoder must be purchased for each TV in the home. The requirement to obtain multiple decoders represents a significant cost penalty for most American homes, which have an average of nearly three televisions per home.

Competition

DBS already faces competition from cable operators and soon will compete with over-the-air digital broadcasters. Ironically, some of the competition from cable will appear in the form of another DBS vendor, namely *Primestar.* Primestar is a DBS owned by the nation's largest cable operators, including TCI and Time Warner. Most marketing for Primestar has been in rural areas so as not to cannibalize cable subscribership.

New DBS competition could come online as well. Canada, Mexico, and some South American countries have been granted orbital slots that could beam service into the United States. At least some of them soon will be auctioning off their spectrum; it is expected that some U.S. companies will bid on that spectrum with the intent of servicing the United States.

Both cable and over-the-air broadcasts have the advantage of local content. Cable also intends to offer interactive services, thereby providing a complete package of services.

Finally, competition might come from Multichannel Multipoint Distribution Services (MMDS) and Local Multipoint Distribution Services (LMDS), both of which are discussed later in this chapter.

Standardization

The vast majority of DBS systems worldwide will use the European DVB standard, with the current exception of DirecTV in the United States. When DirecTV begins operation in Japan, it will use DVB. The prospect of the worldwide use of DVB promises some lowering of costs as multiple manufacturers make common components. But interoperability of set tops among different systems is unlikely because too many variables exist in the DVB specification. In addition, differences occur in conditional access and encryption. In some set tops, these security devices are built into the box, making them tamperproof. In others, the security devices are built into smart cards and consequently can be changed. Interchangeable security was a part of the original DVB specification, but it has not been enforced by service providers who typically have their own ideas about security.

VIEWPOINT

The Role of DBS in Residential Broadband

Because of its one-way capability, DBS is not viewed as an Access Network option for RBB. The role of DBS in this discussion is to further emphasize the importance of television as a market driver and to explore DBS's negative financial impact on other technologies. DBS is instrumental in keeping pricing for cable services low, thereby inhibiting the capability of cable to roll out high-speed services.

Even though it is not a candidate for RBB services, DBS is a spoiler for full-service networks, such as cable, that attempt to provide both data and video services. For the more than 20 percent of homes in the United States that receive DBS, the appeal of other technologies such as xDSL rests largely on the strength of their pull-mode data services because DBS already provides high-quality multichannel broadcast service in such homes.

MULTICHANNEL MULTIPOINT DISTRIBUTION SERVICES (MMDS)

The success of DBS convinced telephone companies and other potential cable competitors that delivering digital video to consumers is a viable business. When such competitors analyzed the competitive issues with DBS, they saw that the lack of local content represents the biggest marketing impact.

Thus, some would-be competitors to DBS sought to improve on it by providing a wireless, multichannel broadband service with local channels called *Multichannel Multipoint Distribution Service (MMDS)*, referred to by DAVIC as Multipoint Video Distribution Systems (MVDS). MMDS grants local over-the-air stations and local advertisers access to digital delivery.

Background

MMDS leverages a microwave transmission technology known as *wireless cable*, which is a microwave technology used to deliver analog cable television service over-the-air to rural areas that could not be served economically by wired cable. Companies with wireless cable services include Heartland, People's Choice, CAI Wireless, and American Telecasting (ATI).

The areas served by wireless cable were too sparsely populated to generate strong revenue growth, so the finances of wireless cable operators have been unimpressive. But the success of DBS and the advent of digital technology (such as MPEG, digital modulation techniques, and advances in semiconductors) changed the perception of microwave from simply a rural delivery system to a system that could be used in urban areas. Telephone companies, particularly PacBell in Southern California, view microwave as a fast-start service to enter video distribution that can compete against cable and DBS. In addition to PacBell, Bell South purchased rights to 1.2 million homes in Florida from ATI for $48 million. This amounts to $40 per home passed, which is a far lower cost than installing a wired network.

Another factor favoring wireless cable is time to market. MMDS can buy time for the telephone companies to offer video service, at least temporarily, until such time as they can justify and install wired networks for broadcast TV. One industry observer noted that MMDS is like saturation bombing by the telephone companies to soften the defenses of the cable operator (with price competition) until such time as the ground attack (wired networking) can be installed. The enhancements to wireless cable to create MMDS include making it digital, offering more channels, and moving it into urban areas.

In 1996, the FCC conducted spectrum auctions for MMDS. The FCC auctions offered 200 MHz in each of the nation's 493 *basic trading areas (BTAs)*. A BTA represents a contiguous geographic market. BTA boundaries are drawn on county lines. The counties are aggregated by considering physical topography, population, newspaper circulation, economic activities, and transportation facilities (such as regional airports, rail hubs, and highways). The BTA concept was licensed by the FCC from Rand McNally.

MMDS frequencies fall in the range of 2.5–2.7 GHz. At these frequencies, the footprint of a transmitter is 30–70 km in radius. Using QAM-64 modulation (not available to early wireless cable operators), the operator can derive an aggregate of roughly a Gb of bandwidth, which is sufficient bandwidth to offer 150–200 channels.

This is comparable channel capacity to DBS. Note that the bit rate available to the MMDS operator is comparable with the bit rate available from DBS systems even though less spectrum is available. This is because MMDS uses more aggressive modulation techniques. DBS has 500 MHz of bandwidth using QPSK modulation (2 b/Hz). MMDS has 200 MHz using QAM-64 modulation (6 b/Hz). After overhead bits and error correction, both DBS and MMDS can achieve nearly 1 Gb of bandwidth.

The auction rules provided no regulations regarding spectrum use. If the operator would rather offer Internet access than TV, or a mix of both, he is free to do so.

MMDS spectrum was auctioned in 1996, but the income from the spectrum auction was a disappointing $216 million for all MMDS BTAs. One can argue that the auction results were hampered because Congress asked that the auction proceed quickly so that the money would be used in the federal budget for the 1997 fiscal year. Had the auction been postponed until the glut of recent bandwidth auctions was clear and until certain technological improvements were available (as a number of FCC staffers and industry observers suggested), the results might have been better. But postponement of the auctions would have postponed the availability of service. There is a clash between Congress's desire to maximize revenue from auction proceedings and the desire to introduce competitive services quickly.

At the time of this writing, the Wireless Cable Association (http://www.wirelesscabl.com/) reported 73 MMDS operators in the United States serving more than 1 million subscribers. Internationally, more than 90 operators serve 5 million subscribers. Various surveys have this number increasing to 3 million to 4 million in the United States by 2001 and 10 million to 13 million internationally as penetration of MMDS moves from rural to urban areas.

MMDS Architecture

The key technical difference between MMDS and DBS is the use of ground-based, or terrestrial microwave rather than geosynchronous satellites. The service difference is delivery of local content, which MMDS achieves by having local production facilities that can insert local over-the-air channels into the national feeds. Figure 6–3 shows a schematic of MMDS service.

National television feeds are delivered by the programmer to a production facility. The feeds can come from other geosynchronous satellite transmission or high-speed wired services, such as fiber-optic networks. Despite what appears to be a good technical fit, there is little current movement to link MMDS with DBS. DBS could provide economic national distribution of programming for resale by MMDS.

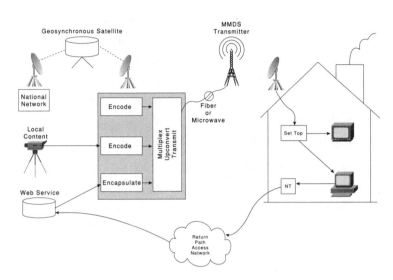

Figure 6–3

MMDS schematic.

Local content and advertising are acquired over-the-air, encoded into MPEG, and multiplexed with the national programming for local distribution to the viewers. MPEG enables digital multiplexing and thus is a key facilitator of MMDS.

Data services also might be received from web content providers. In this case, the information is in digital format but would require additional processing such as encapsulation into MPEG and address resolution before being transmitted.

After the programming mix is determined, composite programming is delivered by satellite or fiber to the MMDS broadcast tower. Generally, the MMDS head end and the MMDS broadcast tower are not co-located because the tower should be placed at a high elevation.

At the receiving site, a small microwave receiving dish a little larger than a DBS dish is mounted outside the home to receive the signals. A decoder presents the TV images to the TV set. Advanced units will be capable of decoding data for PC users.

Return path data is transmitted on another Access Network; telephone networks commonly are used for this purpose. It, for example, is possible to have an RJ-11 telephone jack on the set top box. Consideration is also being given to other wireless networks such as digital PCS and paging networks for return path purposes. However, early rollouts of MMDS generally do not have return path capability.

The range of MMDS primarily is limited by line-of-sight considerations. In relatively flat areas, if the transmitter can be located high enough, the signal can reach over 50 miles. *Pacific Bell Video Services (PBVS)*, for example, currently is rolling out MMDS in Los Angeles and Orange counties in southern California using only two towers. About 75 percent of homes will be able to receive MMDS signals reliably. The remaining 25 percent are impeded by line-of-sight problems.

The Los Angeles MMDS service will have basic service consisting of 41 channels of digital TV and 39 channels of digital audio. Up to 150 channels will be offered for additional fees. Digital video channels are encoded and mixed into MPEG Transport Streams at a digital video head end. The MPEG is transported over an OC-48 fiber link to the MMDS transmitter atop Mount Wilson. At an elevation of more than a mile, Mount Wilson is also the site of most analog TV transmitters in the Los Angeles basin. The signals are modulated onto the MMDS frequencies and beamed out using horizontally polarized microwaves. This signal is transmitted over the entire Los Angeles basin and is strong enough to be received on Catalina Island, 40 km from the coastline and 70 km from the transmitter.

The hilly terrain east of Mount Wilson interrupts line of sight for parts of Orange county. To rectify this, the transmission from Mount Wilson is sent to Mount Modjeska in Orange county, which serves as a relay point to viewers unable to receive Mount Wilson signals. The signals from Mount Modjeska are retransmitted as vertically polarized signals, whereas the original signals from Mount Wilson are transmitted with horizontal polarity.

The horizontally and vertically polarized microwaves enable the two transmitters to have overlapping signal coverage. A dish tuned to one polarity will not receive signals of the other polarity. Polarity is a way to achieve frequency reuse and thereby increase the amount of bandwidth and coverage available systemwide. Divicom supplies the MPEG encoder, and Thomson is the systems integrator.

Southern California is particularly well-suited to MMDS. The Los Angeles basin is relatively flat, its tall buildings are spread apart, and it is densely populated so that a single antenna can reach upward of four million households. This makes infrastructure investment less than $20 per home passed. MMDS is the least expensive infrastructure built for broadcast video with local content.

Because of the availability of telephone return path, MMDS operators are optimistic about their capability to provide a data service very similar to data and cable. Zenith, Hybrid, and General Instruments are leveraging their data and cable TV experience to provide data and MMDS modems using telephone return.

Challenges to MMDS

Despite the low startup costs, a variety of commercial issues remain. Concerns regarding MMDS are reflected in the rather poor showing for MMDS spectrum auctions and the poor financial performance of MMDS operators.

Competition

MMDS faces entrenched competition from DBS and cable TV. Although MMDS has low startup costs, cable is long past the rollout stage. MMDS must compete in pricing and program content, two areas in which it has not been successful to date. Furthermore, over-the-air digital broadcast could further reduce demand for MMDS.

One Way Service

The MMDS spectrum auctioned by the FCC is not approved for return-path transmission. But even if an MMDS return path existed, it would not be practical without fragmenting the return path into smaller cells, such as by using digital cellular technology. In the case of Los Angeles, a single MMDS transmitter can serve four million households. It is infeasible to have return-path service for all converging back to a single central office. To relieve congestion, it would be necessary to segment the MMDS footprint, as cable does for clusters. Consider the return path illustrated in Figure 6–4.

The MMDS transmission boundary encompasses up to a few million households. To provide orderly return-path traffic, each home can transmit back to a PCS base station that covers far fewer users. Each base station would arbitrate return-path congestion and transmit back to the MMDS head end. Each base station would require the equivalent of an access node, which can enforce the Media Access Control (MAC) protocol and format traffic for the return path. Dozens of PCS base stations can exist in an MMDS footprint, probably more than Cable Modem Terminal Servers in the same geographical area. As a result, this is a costly proposition.

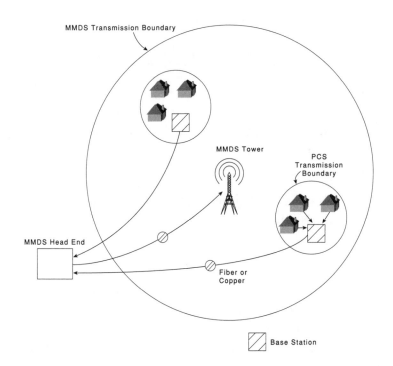

Figure 6–4
Return path using PCS.

PCS service provides a low-bandwidth return path suitable for signaling forward channels but not suited for high-speed return-path applications, such as web hosting on customer sites. Although wireless return is a user-friendly return path, its implementation would remove the cost advantage of MMDS and would require the cooperation of another Access Network carrier.

Industry Financial Condition

Over the past 10 to 20 years, the business of wireless cable operators has been to provide analog TV service in places where wired cable TV didn't want to be. This meant providing service to rural areas, where it was costly to lay cable. Because these areas were often low-income areas, little opportunity to sell lucrative optional programming packages existed.

However, wired cable gradually has expanded its geographical reach even into rural areas. Thus MMDS finds itself competing against wired cable primarily on the basis of price rather than services, because it never had the customer base to offer value-added services.

Now MMDS is faced with finding the investment capital to go digital before DBS takes it all. Capital markets so far have been reluctant to provide funds.

The most likely entrants for digital MMDS are the phone companies, who would like to compete against cable for TV service but do not want to lay coaxial cable. These companies also recognize that xDSL is not optimized for multichannel broadcast television. However, they have not committed to MMDS decisively and, in fact, have pulled back somewhat, as demonstrated in the case of TeleTV. TeleTV was a consortium of PacBell, Nynex, and Bell Atlantic and had plans to produce original programming, enter into programming acquisition agreements, and distribute TV in the major markets in associated territories. TeleTV was greatly downsized after the departure of top management officials when the telephone company supporters decided to wait and see. Today the main effort of the remnant of TeleTV is the Los Angeles rollout of MMDS.

A great deal of the future of digital MMDS in the United States, and probably elsewhere, depends on what happens in Los Angeles. If PBVS can't make a go of it there, then little likelihood exists that Wall Street or anyone else will try it again soon. Competitive two-way services are encroaching on MMDS's viability, in part because some services are able to compete on the basis of capability as well as price, whereas MMDS is largely a price-competitive strategy against cable.

LOCAL MULTIPOINT DISTRIBUTION SERVICES (LMDS)

A more aggressive strategy than MMDS is a new delivery service called *Local Multipoint Distribution Services (LMDS)*, also known in Canada as Local Multipoint Communication Service (LMCS). The major disadvantages of MMDS are the lack of an inband return path and the lack of sufficient bandwidth to surpass cable's channel capacity and offer superior interactive data services. A strong Internet Access Network must have two-way service and enough bandwidth to compete with data and cable.

LMDS is a two-way, high bit rate, wireless service under development by a variety of carriers to solve the return path problem and vastly increase bandwidth. If significant technological hurdles can be overcome, LMDS offers the greatest two-way bit rate of any residential service, wired or wireless, at surprisingly low infrastructure costs.

Background

In January 1991, the FCC granted a pioneer's preference license to CellularVision (http://www.CellularVision.com/) to provide a one-way analog broadcast TV service in Brooklyn, New York. The preference license granted exclusive use of spectrum to more than 3.2 million households. The service used 1 GHz of bandwidth in the 27.5–28.5 GHz frequency band. Prior to CellularVision, this spectrum was granted for point-to-point use, but a waiver was granted so that CellularVision could provide a fixed cellular point-to-multipoint operation for video distribution (wireless cable). CellularVision went public in February 1996. It incorporates components from its strategic partners, Philips Electronics and New Media Electronics (Israel).

The rationale for granting the license was CellularVision's claim that the company had overcome two problems associated with high-frequency wireless transmission. The first problem was how to get sufficient signal strength from high-frequency transmission. CellularVision addressed this by transmitting in relatively small cells. MMDS can transmit 30–70 km in radius. LMDS covers 3–6 km, thus accommodating more and smaller cells.

The reliance on multiple small cells gave rise to the second problem. Small cells create problems of signal interference between adjacent cells. CellularVision attacked this problem by using horizontal and vertical polarization, much like PacBell's solution in Los Angeles. If adjacent cells use different polarity, then they can use the same frequencies without mutual interference. Armed with these innovations, CellularVision obtained its Pioneer's license and became the first and, to date, only LMDS service provider.

CellularVision offers 49 channels of analog, one-way broadcast service. The company also has announced a modest introductory data service of 500 Kbps called CVDN 500, with plans to offer 48 Mbps in 1998. CellularVision has proven that the spectrum is usable, even if commercially the service has not been tremendously successful yet. With the development of wireless digital technologies, renewed interest has been expressed in LMDS as a provider of two-way service.

Regulatory Activity

Interest in LMDS has been building since the FCC issued rules governing the auctioning of bandwidth in May 1997. Auctions for U.S. spectrum are scheduled to begin February 18, 1998 with 139 approved bidders. The rules (FCC Notice of Proposed Rulemaking [NPRM] 97-082) mandate the following frequencies to be used for LMDS service in the United States:

- 27.50–28.35 (850 MHz)

- 29.10–29.25 (150 MHz)

- 31.00–31.30 (300 MHz)

These figures represent a total available bandwidth of 1.3 GHz, which is more than twice the bandwidth of AM/FM radio, television, and cellular telephones combined.

The total of 1.3 GHz is offered in two blocks, called *Auction Block A* and *Auction Block B*. Auction Block A consists of 1.15 GHz and is located at the following frequencies:

- 27.50–28.35 (850 MHz)

- 29.10–29.25 (150 MHz)

- 31.075–31.225 (150 MHz)

Auction Block B is located at the following frequencies:

- 31.000–31.075 (75 MHz)

- 31.225–31.300 (75 MHz)

The segmentation of bandwidth creates greater separation between transmitting and receiving frequencies, eliminating the need for filters and ensuring efficient use of LMDS's spectrum.

Quoting from the NPRM, the following list details the major elements of the LMDS auction rules:

- The LMDS spectrum will be licensed by basic trading areas (BTAs), for a total of 984 authorizations and 1,300 MHz of spectrum.

- Two licenses, one for 1,150 MHz and one for 150 MHz, will be awarded for each BTA.

- All licensees will be permitted to disaggregate and partition their licenses.

- The number of licenses a given entity may acquire is not limited.

- Incumbent local exchange carriers and cable companies may not obtain in-region 1,150 MHz licenses for three years.

- LMDS may be provided on a common carrier or on a non-common carrier basis, or both. *Common carrier* status means that the network provider would provide resale rights openly on a nondiscriminatory basis to content providers. *Non-common carrier* status means that the network provider would provide content itself and not be required to resell the raw transport service.

- Licensees will be required to provide "substantial service" in their service areas within 10 years.

- Incumbents in the 31 GHz band will be able to continue their operations but will receive protection from LMDS operations only in the outer 75 MHz (31.0–31.075 and 31.225–31.300) of the band. Incumbents will be given 75 days from the date of the publication of this item in the Federal Register to apply to modify their licenses to operate in the outer 75 MHz of the band, and an additional 18 months to implement those modifications.

• Bidding credits and installment payment plans will be available to small businesses and entities with average annual gross revenues of not more than $75 million.

The provision that prevents telephone companies and cable operators from obtaining in-region licenses for three years is based on the fear that if these groups were to win spectrum, they simply would sit on it for the 10 years allotted before they are required to provide substantial service. Taking the spectrum out of circulation in this way would reduce competition for the existing services. The historic reluctance of local exchange carriers to invade each other's territory, coupled with this provision, will have the effect of keeping them out of the auctions entirely. At the time of this writing, the provision is being challenged in court by LECs who want to enter the LMDS auctions. The effect of legal challenges could be the postponement of the auctions until 1998.

Long-distance carriers are not precluded from bidding on spectrum and are likely to be active in the auctions because LMDS is a logical way of bypassing local exchange carriers to provide broadband service. MCI, for example, is partnered with British Telecom, which in turn has been running trials of wireless cable since the 1980s. MCI can leverage British Telecom's experience with wireless networks to offer end-to-end broadband service and therefore is a likely bidder in the auctions. The long-distance carriers are further motivated to bid because of the provision that no limit be put on the number of territories owned. It is possible for a single long-distance carrier to have a seamless national broadband network.

LMDS as a Telco Business Strategy

The interexchange carriers (IXCs) would like to have inexpensive access to dry copper pairs from local exchange carriers (LECs) so as to compete against LECs for voice service. That way, the IXCs could connect to the LECs in the same way in which they have since 1984 rather than rolling out all those LMDS towers. Currently, the IXCs are not satisfied with dry copper tariffs.

Winning an LMDS license would be a bargaining chip for IXCs in their negotiations with the LECs for dry copper pricing. The IXCs could use LMDS as a threat to obtain better tariffing. If they don't get the tariffing they like, IXCs could bypass the LEC using LMDS. If they do get the tariffing they want, LMDS rollout might take a while.

No restrictions exist as to how the awardee uses the bandwidth, so bandwidth can be subdivided in any manner the carrier sees fit. If an LMDS carrier had 1,150 MHz of bandwidth, for example, it would be possible to use 500 MHz for broadcast TV, 50 MHz for local broadcast, 300 MHz for forward data services, and 300 MHz of upstream data. Using only the relatively robust QPSK modulation, this bandwidth can provide the following:

- All the broadcast channels of DBS (500 MHz)

- All local over-the-air channels (50 MHz)

- Up to 1 Gb of full duplex data service (600 MHz)

In other words, the potential exists to offer more TV than satellite and more data than cable. This frequency plan is just one example of how a carrier could choose to offer service. Other carriers might choose to segment their frequencies differently and would be permitted to do so under FCC rules.

Spectrum rules and allocation are different elsewhere. In Canada, the service is LMDS spectrum and is allocated between 27.35 GHz and 28.35 GHz. Another 2 GHz is being held for future services. In South America, Asia, and some parts of Europe, the spectrum is 27.50–29.50 GHz.

LMDS Architecture

LMDS is a small cell technology, with each cell about 3–6 km in radius. Small cells coupled with two-way transmission create a different set of architectural problems than MMDS. Figure 6–5 shows a schematic of LMDS service.

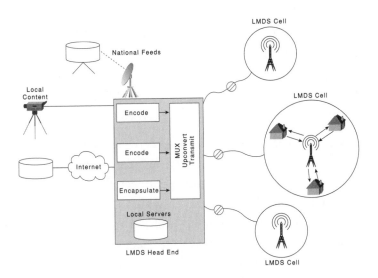

Figure 6–5
LMDS schematic.

Content acquisition at the LMDS head end functions similarly to MMDS. National television feeds are delivered by the programmer to a production facility. In many cases, these national feeds come from DBS, but the feeds also can come from other geosynchronous satellite transmission or high-speed wired services such as fiber-optic networks.

Local content and advertising are acquired over-the-air, encoded into MPEG, and multiplexed with the national programming for local distribution. As in the case of MMDS, MPEG is an important facilitator of LMDS because it enables digital multiplexing.

Data services received from web content providers are already in digital format but would need additional processing, such as encapsulation into MPEG and address resolution before being transmitted.

The program mix is delivered by satellite or fiber to the LMDS broadcast tower. Generally, the LMDS head end and the LMDS broadcast tower are not co-located, because the head-end production facilities would be shared among several towers.

An LMDS transmitter tower is erected in the neighborhood, and traffic is broadcast to consumers using QPSK modulation with FEC. It is possible to use QAM modulation, but QPSK is chosen because it is more robust than QAM 16 or QAM 64 and because bandwidth is so plentiful that spectral efficiency is not an issue.

As shown in Figure 6–6, consumers receive the signal on a small dish about the size of a DBS dish or a flat-plate antenna. The dish is mounted outside the home and is connected by cable to a set-top converter, much the same way in which DBS connections are made. The signal is demodulated and fed to a decoder. Unlike DBS, LMDS is capable of two-way service, so both TV sets and PCs must be connected to the satellite dish. Furthermore, a two-way home networking capability must be supported instead of just the simple broadcast scheme of DBS. This places new requirements on home equipment, which will be explored in the next chapter on home networking.

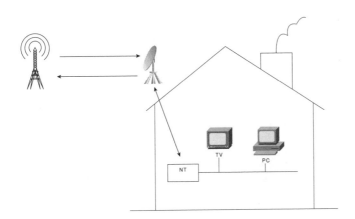

Figure 6–6

LMDS in the home.

In the return path, the customer transmits to the carrier using the same dish with QPSK modulation. A Media Access Control (MAC) protocol is required because the residences in the coverage area share the return spectrum. In Figure 6–5, for example, all three consumers in the middle LMDS broadcast boundary receive broadcast TV and Internet traffic on the same frequency, say 27.5–28.35 GHz. Likewise, all consumers share the same frequencies for upstream transmission, say 29.10–29.25 GHz. In this case, all three users would contend for upstream bandwidth.

Architecturally, LMDS looks very much like cable TV. Cable TV clusters serve 500–2,000 homes; LMDS cells serve 2,000–6,000 homes. It is not surprising then that the MAC protocol is similar to cable TV, as are the ASICs for the customer premises modulators and demodulators. Upstream users request data slots on a contention basis. After slots are granted, the sender transmits in those slots free of contention. Ranging and power-level controls are also required, as in the case of cable.

One way to increase the amount of frequency available is to use sectorized antennas using space division multiple access (SDMA), which broadcasts and receives to and from a pie-shaped segment of the transmission area. SDMA is a frequency reuse technique that minimizes congestion, much like cellular telephony. This technology is illustrated in Figure 6–7.

Figure 6–7
LMDS sectorized.

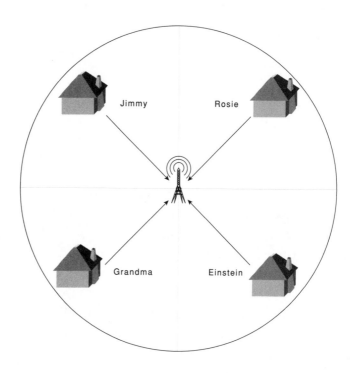

Jimmy, Rosie, Grandma, and Einstein are all transmitting on the same frequencies back to the head end. If no sectors existed, then the four households would contend for common spectrum. But sectorization enables the same spectrum to be reused in each pie-shaped area. In Figure 6–7, it is possible for each household to transmit on the same frequency in the upstream direction at the same time because they use different sectors. This increases the amount of frequency available for the return path by a factor of 4, given 90-degree sectorization.

An LMDS tower and transmitter are projected to cost about $500,000 to $1 million each. CellularVision's layout has 12 transmitters in place in Brooklyn and Queens, New York, providing coverage to 28 square miles or 72 square km, which is about 1 transmitter per 6 square km. Assuming that each square kilometer contains 1,000 residences, 6,000 homes would be passed per transmitter. This brings capital cost per home passed to $100–$150. In less dense areas, the cost per home passed increases, but not necessarily linearly with respect to density. This is because it is possible to have fewer transmitters. Local conditions (such as rain and foliage) and the expected customer take rate all have a role in planning cell layout.

Challenges to LMDS

Despite the advantages of LMDS compared to wired networks—including large amounts of bandwidth and relatively low startup costs—the fact remains that only one LMDS provider exists today and that relatively few vendors are developing the product. Texas Instruments, an early LMDS provider with pilots in North and South America, recently sold its LMDS business to a European company, Bosch Telecom, the second-largest microwave provider in Europe, behind Siemens. One reason for the limited commercial development is that the challenges to LMDS are daunting.

Signal Quality

As with microwaves in a microwave oven, LMDS microwaves react with water droplets in the air or in foliage, causing LMDS to lose signal strength. At 30 GHz, the wavelength of the LMDS signal is about a millimeter in length, whereas MMDS wavelength is 11 times longer and DBS is longer still. With that small wavelength, LMDS signal quality can be attenuated by something as small as a raindrop or a drop of tree sap. LMDS also requires clear line of sight. Because of these problems, LMDS is usable only in relatively flat areas with relatively little foliage. It might be possible to overcome foliage problems with more efficient modulation and antennas, but doing so would incur higher transmitter costs and possible regulatory problems. Furthermore, such innovations do not help return path problems at all.

Technical innovations in development are intended to overcome challenges to signal quality. The idea is to exploit multipathing to recover the original signal.

Figure 6–8 shows a transmitter and a consumer with foliage between them. The multipath technique for recovering the signal combines the weak signal received through the foliage with signals that bounced off other surfaces. The receiver looks at the multiple signals, accounts for the respective delays of each path, and derives the original signal. The technique works as long as the delay spread is not too great. The *delay spread* is the time difference between the first and last path received. A delay spread of several microseconds can be accommodated.

Figure 6–8
The multipath technique.

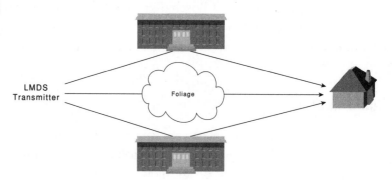

This technique makes LMDS preferable to MMDS in environments that contain lots of solid objects (such as buildings) that cause signals to bounce. MMDS cannot use the multipath technique without changing the underlying modulation and using multiple antennas at one end of the link. The underlying question is the cost effectiveness of multiple antennas. LMDS microwaves scatter more efficiently, although the scattering efficiency depends on the reflectivity, texture, and porousness of the surface. Because of this multipath correlation technique, a few proponents assert that LMDS is capable of operating without having a direct line of sight with the receiver. Still, most experts are skeptical of this claim.

One potential drawback with the multipath technique is its effect on polarity. Remember that combining vertical and horizontal polarity provides frequency reuse between adjoining cells. With substantial multipathing, it is likely that polarity cannot be maintained as signals bounce off surfaces.

The whole issue of getting good signal integrity is still the subject of lively debate. But field tests in the United States and Brazil (where it rains a lot) have given encouragement that these problems are on their way to being solved. In a Brazilian test run in 1996, Texas Instruments reported that equipment was able to operate at a radius of 6.5 km. Vebacom in Germany has error-free microwave links in the 23 GHz band that can cover 10 km.

But doubts about signal strength persist on Wall Street, where LMDS is viewed primarily as a solution for flat, high-density, low-rainfall, low-foliage areas.

Cell Size

Smaller cell size increases infrastructure cost (necessitating more towers and transmitters) but improves coverage and reliability to consumers. The following list details the main engineering factors that can increase cell size:

- Increased transmission and receiver antenna height

- Increased power

- Overlapping cells versus single transmitter cells

Increasing the height of the transmission and reception antennas greatly improves reception by improving line of sight, mainly by transmitting over trees. The downside is aesthetic and cost. Operators will try to construct towers as high as possible consistent with local ordinances and tower costs.

In rainy weather, the operator might increase transmission power slightly to get more energy through the network. No adaptive transmit power exists; instead, power levels are set during trials to provide the required coverage. This increases transmitter cost and can cause interference.

Another technique for better reception is to have overlapping cells. In a single transmitter cell, all homes must send and receive to a single tower. Hewlett-Packard found that a single transmitter would reach only slightly more than 60 percent of the homes in a cell due to line-of-sight problems. With overlapping cells, a given household could be within range of two or more towers and could select the tower that provides the best signal strength by measuring power levels. Studies by Hewlett-Packard have shown that overlapping cells can substantially increase penetration rate to nearly 100 percent.

The challenge for LMDS operators will be creating economical deployment strategies. As a point of contrast, the design of cell size and the location for LMDS is unlike that of cellular telephone. Cellular telephone is a wireless system often used by consumers while they are in transit. For such a system, it is important to start with large cells and gradually split them into smaller cells as usage grows to minimize handoffs. A *handoff* is the process of transferring the responsibility for a cellular voice call from one base station to another as the caller moves out of range of the original cell. During handoffs, voice calls can go silent as the cellular network sorts out which base station is responsible for maintaining the call.

No handoff problem exists in the LMDS case, because none of the subscribers are in motion. This means that LMDS operators must start with small cells to maximize coverage and reliability. Such a process is costly and changes deployment strategies from the cellular model. Although broad coverage was the key to success in a cellular network, identifying and serving areas densely populated with potential customers is key to LMDS. Instead of segmenting big cells into small cells (such as with cellular operators), LMDS operators must start with small, well-placed cells and add more small cells as service levels increase.

Carrier Buildout Costs

Owing to its small cells, LMDS requires many more transmission towers than MMDS. Because MMDS has about 10 times the transmission radius of LMDS, LMDS requires about 100 times more towers to cover the same area. In addition to construction costs, delays are associated with site selection, easements, permits, and negotiation of rentals. The result is that LMDS is significantly more expensive and time-consuming to rollout than MMDS.

Cost of Consumer Equipment

Unlike DBS antennas, LMDS dishes can transmit as well as receive. This increases cost for consumer equipment compared to DBS. Furthermore, installation costs are associated with LMDS; unlike DBS, for instance, an installer would be needed to align the antenna. Many analysts have concluded that DBS is the upper limit of what consumers will pay for digital TV equipment, and LMDS could push costs beyond that.

AT ISSUE

Fragmented Area Coverage

Forthcoming LMDS spectrum auctions raise the question of another potential challenge. The current approach to spectrum auctioning in the United States is to conduct separate auctions for different geographical areas, namely the BTAs. This has the tendency to fragment the country into multiple potential providers. Such an approach was taken for cellular telephone and PCS. Benefits of regional auctions include increasing the number of competitors nationwide and increasing auction incomes.

Canada is instituting another approach. There, a nationwide license was issued to MaxLink Communications, Inc. by the Canadian Department of Industry (Industry Canada) for nearly half the population of Canada. This approach offers incentives to the providers because they have a bigger market to exploit with a single license. It remains to be seen if the checkerboard procedure or the national licensing approach serves best to stimulate broadband rollout.

Finances

To date, CellularVision is the only U.S. company providing service for LMDS, and it does so partly on the strength of the preference the government shows. So far, CellularVision's stock performance has languished behind even cable operators. Wall Street remains concerned about the capability of LMDS to solve the problems of rain, foliage, consumer demand, and competition from DBS, cable, and (soon) over-the-air digital broadcast for distribution of video. It appears that for LMDS to succeed, private capital—namely the retained earnings of major telephone companies—is required to give this industry a boost.

In addition to LMDS providers, there are other providers of high frequency, wireless, terrestrial services. Chief among these are Teligent (using frequencies in the 24 GHz range) and Winstar (in the 38 GHz range). Both plan to offer services similar to LMDS.

LOW EARTH ORBIT SATELLITES (LEO)

This review of wireless technologies concludes with a discussion of *Low Earth Orbit (LEO)* satellites. Though it is unlikely that the bandwidth of these systems will exceed 1–2 Mb, their potential role in RBB is to provide a ubiquitous return path for other one-way technologies. Furthermore, analysts of potential RBB networks must calibrate the impact LEOs will have on the financial prospects of voice carriers. If, for example, LEOs are highly successful in providing mobile voice and data service, they could fragment the residential market and negatively impact the development of other high-speed services.

Background

Unlike their geosynchronous brethren, which are at 35,800 km above the equator, LEO satellites orbit the planet at low altitudes. Depending on the system, the altitude of these orbits ranges from 780–1,400 km, which is above Earth's atmosphere but below the Van Allen radiation belt. At these altitudes, a LEO satellite is in view of 2–4 percent of Earth's surface, which means that the footprint of any given LEO satellite is 4,000–6,000 km in diameter, depending on altitude.

The low altitude provides two advantages for the consumer. First, latency is short. Phone calls through GEOs incur a 239 ms one-way delay; a round trip delay is therefore nearly half a second. This causes annoying pauses during voice conversations.

One-way latency for LEOs is about 6 ms. The second advantage is lower power consumption. LEOs travel so low that a handset requires very little power to reach it, as compared with a GEO.

Given these advantages and the recent advances in satellite launch technology, a number of companies have initiated development programs. Among these are Globalstar, Motorola Iridium, Motorola Celestri, and Teledesic. The objective of Globalstar and Iridium is to provide ubiquitous voice service. Celestri and Teledesic have plans to provide bandwidth up to 2 Mb.

LEO Architecture

Because LEOs fly at low orbit, they move with respect to the Earth. This means that as a user on the ground communicates with the LEO, the LEO passes over the horizon and communication stops. To maintain the session, the original LEO must handoff the session to the following satellite. By succeeding handoffs, the session can be maintained. The frequency of handoffs is determined largely by the distance of the LEO from Earth, which determines the footprint size of the LEO. The lower the satellite, the faster it moves with respect to Earth, creating more frequent handoffs. There are possibly hundreds of satellites communicating with each other, handing off user sessions. The original Teledesic system had nearly 900 satellites until a recent redesign reduced that number to 288. Lower orbits increase the number of satellites, which is an important component of system cost.

Two different approaches are being taken regarding switching. Globalstar takes a "bent pipe" approach, in which traffic is sent from the user up to the LEO. The LEO handles minimal processing of the bits and returns the traffic to the ground as soon as possible. All switching decisions are made on the ground.

Iridium and Teledesic take the approach of switching in the sky. Each satellite contains an ATM or packet switch to forward packets to other satellites. When the final satellite is reached, that satellite beams down directly to the end user or the ground station. Switching in the sky minimizes the cost of ground stations and handsets and also reduces the requirements for ground switching.

Globalstar and Iridium expect to offer services in the second half of 1998. Iridium already has launched some of its satellites from the Balkanour facility in Kazakhstan. Teledesic has not announced a service date, but the time period is expected to be after 2001. Table 6–2 summarizes the features of these three LEO competitors

Table 6–2 *Features of LEO Competitors*

	Globalstar	Iridium	Teledesic
Downlink bandwidth	2.483–2.500 GHz S-band	1.616–1,625 GHz L-band	19.3–19.6 GHz Ka-band
Uplink bandwidth	1.610–1.626 GHz L-band	1.616–1,625 GHz L-band	29.1–29.4 GHz Ka-band
Modulation scheme	QPSK	QPSK/CDMA	N/A
Switching	On the ground; uses bent pipe approach	In the satellite	In the satellite
Satellites, with spares	56	72	288

Table 6–2   *Features of LEO Competitors, Continued*

	Globalstar	**Iridium**	**Teledesic**
Maximum bit rate/ session	9,600 bps	4,800 bps	16 KBps (voice) 2 Mbps (data)
Investors	Publicly traded (Nasdaq: gstrf)	Motorola, Lockheed, Martin, Sprint, and carriers worldwide	Bill Gates and Craig McCaw

Challenges to LEOs

Not surprisingly, the futuristic and spectacular concepts behind LEOs face some down-to-earth problems.

Telephone Interconnection

Voice service will require cooperation with land-based telephone companies. If a LEO customer were to initiate a phone call to a PacBell user, at some point PacBell must agree to accept the call. Similarly, LEO operators must sign settlements and interexchange agreements with telephone companies worldwide.

Spectrum Use

Conflict arose between Teledesic and LMDS operators about use of the 29.1 GHz spectrum (Ka-band). Eventually, discussions with the FCC resolved the issue. However, LEOs are global systems, so spectrum-utilization problems could recur in other countries that use the same spectrum. Little use of Ka-band occurs worldwide, but full global coverage requires spectrum agreements among S-, L-, and Ka-bands. In addition, radio astronomers are concerned about LEO impacts on their frequency measurements.

Launch Capacity

Finding satellite launch space will be a challenge. More than 1,700 satellite launches are planned for the next 10 years for uses other than LEOs. LEOs will require the launch of more than 400 satellites over the next five to seven years. Considering that there were 22 rocket launches worldwide in 1996—which placed 29 satellites in orbit—the launch business needs to add capacity quickly to meet launch demand. To make things worse, about 10 percent of launches fail. Insurance premiums in some cases are nearly 25 percent of payload value. Incidentally, only about 30 percent of satellite launches are handled by the United States. About 60 percent of the world's launches are made by the European space consortium, Ariane. Others in the launch business are China and Russia.

Cost Issues

In general, there is widespread skepticism about lifecycle costs for LEOs. The original estimate provided by Teledesic was $9 billion to launch an 840-satellite constellation. Independent industry observers made much higher estimates. Over the past year, Teledesic reduced the number of satellites from nearly 900 down to 280 by increasing altitude of the orbits. The following list details some of the cost concerns of particular relevance for LEOs:

- Capital costs for satellite development, construction, and launch. For example, data communications equipment must be modified for operation in space to accommodate environmental factors such as temperature and radiation. These costs are unknown at present.

- Technical innovations required, such as satellite-to-satellite communications at multimegabit rates.

- Continuing requirement for ground stations and associated settlement costs.

- Damage due to solar activity and small projectiles.

- Development of new handsets and customer terminals.

The concept of LEOs exhibits a boldness not found except in science fiction, with many carriers buying into the concept. LEOs offer global roaming, a simple worldwide dialing plan, and instant voice infrastructure for developing countries. In addition, Teledesic is offering megabit service. The impact on RBB is simply that every one-way technology has the possibility of a low-latency return path, which can be used everywhere in the world. More broadly, if successful, LEOs can substantially alter how we think about global communications.

EFFECT OF SPECTRUM EFFICIENCY ON AUCTIONS

Earlier, this chapter noted that LMDS service will provide more than twice the bandwidth of AM/FM radio, television, and cellular telephone combined. The spectrum efficiencies afforded by wireless technologies raise some important issues for the general auctioning of spectrum.

The U.S. taxpayer has benefited greatly from the auctioning of the nation's airwaves. The aggregate proceeds from 15 FCC auctions account for more than $23 billion for the U.S. Treasury (see `http://www.fcc.gov/wtb/auctions.html` for details). Congressional budgeters, pleasantly surprised by the initial reaction to PCS auctions in the mid-1990s, look upon the auctions as a relatively painless way of filling federal coffers. That's the good news.

The bad news is that it's possible to have too much of a good thing. The new reality is that spectrum is being used more efficiently, so less of it is needed and auctions stand to generate less income. More aggressive modulation techniques, digital compression, advanced coding techniques, and forward-error correction all conspire to deliver more service with less bandwidth. Digital voice will be able to compress more voice calls in the same bandwidth as today's analog cellular telephone, some say by a factor of 10, while others say a factor of three. Either way, a lot more voice capacity will exist in the air when digital telephony replaces analog voice.

Similarly, the return and auctioning of analog TV spectrum will have a huge impact on spectrum pricing. Analog TV spectrum is truly prime real estate. It is low frequency, so it can travel long distances and can be modulated aggressively. When these frequencies become available over the next decade, the value of existing licenses could plummet, and the Congressional budget will lose its golden egg.

The value of spectrum allocation is based on how much of it you have, as well as its location in the overall spectrum. The more bandwidth, the better; the lower the frequency, the better. Table 6–3 compares these criteria for a number of services. By examining the criteria of bandwidth and location, it seems as though the soon-to-be-freed broadcast bandwidth could be the most valuable of all.

Table 6–3 *Bandwidth and Location of Various Networks*

	Bandwidth (MHz)	Location (MHz)
Cellular Telephone	25	850
Personal Computer Service	120	1,900
Wireless Communication Service	30	2,300

Table 6–3   *Bandwidth and Location of Various Networks, Continued*

	Bandwidth (MHz)	Location (MHz)
MMDS	200	2,500
DBS	500	12,200
LEOs, Globalstar, Teledesic	16 17 300 300	1,610 2,483 19,300 29,100
LMDS	1,150	27,500
Broadcast TV	348	54–806

SUMMARY

Multiple wireless options exist that potentially can support RBB services. The services discussed in this chapter—DBS, MMDS, LMDS, and LEO—overlap somewhat in functionality but have enough differences to attract their particular segment of users. Table 6–4 compares features among these options.

Table 6–4   *Feature Comparison of Wireless Access Networks*

Feature	DBS	MMDS	LMDS
Bandwidth	500 MHz	200 MHz	1,150 MHz (A-band) 150 MHz (B-band)
Spectrum range	12.2–12.7 GHz	2.5–2.7 GHz	27.50–31.30 GHz
Modulation scheme	QPSK	QAM 64	QPSK
Downstream bit rate capacity	Roughly 1 Gb	Roughly 1 Gb	Roughly 1.5 Gb
Return path	Telco return, digital cellular, or xDSL	Telco return, digital cellular, or xDSL	LMDS return, up to carrier decision—assume 150 MHz of return path

Table 6–4   *Feature Comparison of Wireless Access Networks, Continued*

Feature	DBS	MMDS	LMDS
Upstream bit rate capacity	Narrowband	Narrowband	300 Mb
Footprint	Nationwide	40–60 km radius	3–6 km radius
Homes passed per cell	Nationwide	Greater than 1 million	A few thousand
U.S. service providers	DirecTV, Echostar, and Primestar	PacBell Video, American Telecast, Heartland, CAI Wireless, and People's Choice	CellularVision, and bidders for LMDS spectrum TBD
Tower height	N/A	60–70 m	20–30 m
Rainfall attenuation	Negligible	Negligible	Possible showstopper
Challenges	Lack of local content for TV Lack of a return path	Lack of a return path Weak programming packages to date DBS and cable competition	Most costly wireless rollout; many towers Rain, foliage Requires dense residential use; urban areas only
Facilitators	Successful broadcast TV story with good content	Lowest cost infrastructure Accommodates local content TV	Best two-way story Most bandwidth in both directions Local content TV

Table 6–5 summarizes the ways in which various features of wireless services act to challenge or facilitate their prospects for succeeding as an RBB network.

Table 6–5 *Challenges and Facilitators for Wireless as an RBB Network*

Issue	Challenge	Facilitator
Media	Difficult transmission characteristics exist. Line of sight is required. Multipathing and interference present continuing problems.	Minimal installation and maintenance concerns exist. Fast service rollout will be available once the target market area is decided.
Spectrum	Different spectrum allocations in different countries require coordination for global service and equipment development.	Spectrum is plentiful due to recent modulation and compression techniques and newly usable high frequencies.
Local content service	Service is not available for DBS until the resolution of spot beaming and copyright issues.	Services provide market advantages to LMDS and MMDS for TV service.
Noise mitigation	A shared topology means that a single noise source can adversely affect multiple households.	Similar to cable, sophisticated noise-mitigation and problem-isolation techniques are under development; some cable techniques can be repurposed.
Signaling	New methods are required to remotely control the subscriber unit for return path use.	Subscribers are always connected, so there is no need for a call setup process for data or video, which reduces signaling requirements compared with telephone services.
Data-handling capabilities	Implementation of a return path for DBS and MMDS is required; adding it is feasible but expensive.	Technology maps well to Internet data; high-speed forward and low-speed return paths closely map to web access and push-mode data. LMDS will offer the highest two-way bit rate of any residential service if technological challenges are solved.
Regulation	Continued postponement of LMDS auctions exists in the United States. New regulations are possible on DBS. Copyright issues must be resolved.	No restrictions exist as to how bandwidth is used by carrier.

Table 6–5   *Challenges and Facilitators for Wireless as an RBB Network, Continued*

Issue	Challenge	Facilitator
Market	Capabilities offer the highest bit rates. Cost advantages might yield lower consumer prices.	Being late to market with product development could hinder development.
Business/Financial	Perception of continuing transmission difficulties inhibits investment. A poor investment history exists for current wireless cable service providers.	Major telcos appear willing to invest in part because it provides bypass of LECs. The strongest competitive options go up against cable for new market entrants who want to provide data and video services.

This chapter completes the survey—begun in Chapter 3, "Cable TV Networks"—of specific networks that could provide RBB service. The next chapter turns to the home networking, the point at which the carrier network's responsibility ends and the resident's begins.

REFERENCES

Articles

"Dollar Days: Sale of FCC Licenses In Several States Nets Budget Pocket Change." *Wall Street Journal*, June 3, 1997.

Honcharenko, Walter, Jan Kruys, David Lee, and Nitin Shah. "Broadband Wireless Access." *IEEE Communications*, January 1997.

Internet Resources

http://www.ajs2.com/lmds
(Lots of LMDS information.)

http://www.broadband-wireless.org
(Lots of broadband wireless information.)

http://www.businessweek.com/1997/04/b35118.htm
(DBS and LMDS statistics.)

http://www.cais.com/
(Wireless Cable Association.)

http://www.CellularVision.com/
(CellularVision provides wireless analog broadcast service to a tough audience—Brooklyn.)

http://www.directv.com
(The home page for DirecTV.)

http://www.dbsdish.com/dbs/a0.html
(DBS FAQ from the DBS Connection.)

http://www.fcc.gov/
(The Federal Trade Commission weighs in on LMDS.)

http://www.fcc.gov/Bureaus/Wireless/Orders/1997/fcc97082.txt
(The word from the FCC on LMDS, as well as auction rules and technology background.)

http://www.fcc.gov/oet/faqs/pioneerfaqs.html
(Rules on getting a Pioneer's preference license from the FCC.)

http://www.fcc.gov/wtb/auctions.html
(The FCC's auctions home page, which covers what was auctioned, how much was raised, and who the winning bidders were.)

http://www.fcc.gov/wtb/auctions/lmds/nr970011.txt
(Press release accompanying FCC auction announcement.)

http://www.idt.unit.no/~torwi/overview.html
(Tor Wisloff's excellent overview of big LEOs.)

http://www.internettelephony.com/archive/9.30.96/
CoverStory/coverstory.html
(LMDS brief.)

http://www.itscorp.com/tech/a2wayv12.html
(ADC talks about two-way MMDS.)

http://www.law.cornell.edu/uscode/17/119.html
(Satellite Home Viewing Act, paragraph 119, which discusses royalties and broadcast restrictions on DBS.)

http://www.multichannel.com/dbs.htm#dbshome
(Pointers to DBS web sites.)

http://www.ntia.doc.gov:80/osmhome/osmhome.html
(Spectrum allocation information from the U.S. Department of Commerce/Office of Spectrum Management/NTIA.)

http://www.pchoicetv.com/TECH/TECH.HTM
(Overview of MMDS.)

http://www.refreq.com/braddye/index.html
(Informal site for paging information.)

http://www.wirelesscabl.com/
(Home of the Wireless Cable Association.)

http://www.wlana.com//index.html
(Home of the Wireless LAN Alliance.)

CHAPTER 7

In-Home Networks

The function of wired and wireless Access Networks is to transfer digital traffic to and from the home. The Access Network connection to the home is made at the Network Termination (NT), which demarcates the legal and business responsibility of the carrier. From that point on, traffic must be transferred to the terminal equipment over a network under the control of the resident. In addition to moving to and from the carrier, traffic also can move among various terminal equipment in the home without entering the Access Network. Moving the bits around the home is the function of network infrastructure and equipment that comprise the Home Network.

The Home Network is critical for distribution of data to the various customer premise devices that will be attached—TVs, PCs, stereos, alarms, and so on. It must support multiple data types with high security, easy configuration, low cost, and negligible maintenance. This chapter presents requirements, issues, and alternatives for networking in the home, and motivates the need for IEEE 1394 support and a home gateway architecture.

EXISTING HOME NETWORKS AND SERVICES

Home Networks are not a new invention. For years, telephone companies, cable companies, and electric utilities have installed wiring in homes for the transmission of voice, analog television service, and electric power. In addition, many newer homes have networks

381

for environmental control systems to regulate air conditioning and heating. Some Home Networks exist for intercoms, entertainment systems, and home security systems. The installation of these services requires the service provider to perform some form of home wiring.

In addition to these networking industries, the *consumer electronics (CE)* industry is expected to play a role in the Home Network. The CE industry currently distributes game players and television sets. Game players are a source of demand for home networking bandwidth and, perhaps more interestingly, have the processing power and operating system support to evolve into entities with a networking role. The strengths of the CE industry are its history in shipping complex pieces of electronic equipment at low margins, its distribution channel, and its knowledge of consumer preferences in entertainment and branding. Its weaknesses are lack of familiarity with systems integration and lack of carrier and networking experience. CE vendors clearly have a desire to protect their turf in the home and seem to be willing to partner with carriers, system integrators, and networking vendors to complement their strengths.

Based on experience with existing networks and the desires of commercial interests, basic questions come to mind. Can the existing Home Networks be repurposed for high-speed data networking to support Internet service and/or digital television? If homes must be rewired, how much does the market for RBB shrink? How do various commercial interests best position their products?

Various industries and the ad hoc and dejure standards organizations that support them view the home networking problem differently. Some advocate wired approaches, although others champion wireless approaches. Other potential solutions include systems that have the power to support video and others that don't, as well as systems that use existing wiring and others that require installation of new wire. No clear leader is emerging from the array of possible solutions. Product development lags when compared with developments in Access Networks. The status of standards and the production of components and finished products trails HFC and xDSL by a year or more.

The objective of the rest of this chapter is to describe requirements, issues, and alternatives for home networking and thereby frame the discussion for readers wanting to pursue this complex topic further.

HOME NETWORK REQUIREMENTS

Identifying Home Network requirements is difficult given the lack of consensus about what services, media, and administrative responsibilities define a Home Network. Thus, the network requirements discussed in this section are not universal; they vary somewhat given specific implementation alternatives. The section begins with connectivity options: different services and levels of interconnectivity among them that can constitute the Home Network. The chapter then examines specific technical requirements associated with some or all the options.

Connectivity Requirements

Figure 7–1 depicts the various connectivity options for in-home traffic.

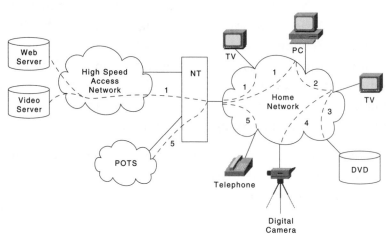

Case 1 shows the required traffic from the NT to various devices within the home. Multiple devices on the same Home Network create a challenge when they access the same or different carriers simultaneously. The TV viewer, for example, might be viewing interactive TV, while the PC user might be active in a web-browsing session. The resulting challenges include address resolution, bandwidth allocation inside the home, and signaling.

In Case 2, traffic is moved from the PC for display on the television set. Currently a TV set costs much less than a PC monitor for comparable screen size (a Sony 17-inch TV costs a lot less than a Sony 17-inch monitor). Thus, using a TV monitor instead of a PC monitor can reduce the total cost of home computing, but doing so requires a high-speed interface between the PC and the monitor.

In Case 3, a viewer wants to watch a movie that is stored on a high-capacity storage device, such as the new *Digital Video Disc (DVD)* currently being promoted by CE vendors. DVDs have the storage capacity for an entire movie in high-definition with five-channel surround sound on a single disc. DVDs also store computer data. Initial versions hold 4.7 GB, far exceeding today's CD-ROM. A DVD can even read today's 600 MB CD-ROMs and are expected to be the dominant form of mass storage for the home as well as business. Access to these devices most likely will be a requirement of Home Networks.

A variation on Case 3 is multiple viewers watching the same show in different rooms, suggesting some form of broadcast or multicast within the home.

In Case 4, the TV displays output from a digital camera in real time. A challenge here is that the camera is unlikely to have built-in memory and processing capability. Therefore, another home device, such as a set-top box, is required to provide signaling support on behalf of the camera so that the camera can direct its output to the correct display device.

In Case 5, narrowband devices such as telephones, meters, and home automation controls could require access to an external carrier network. Though these devices require little bandwidth, they do have high requirements for availability, more so than video or computing. In addition, narrowband devises present new problems of address resolution and network management.

A number of technical requirements emerge for consideration given these five connectivity options and are discussed next. In addition, connectivity raises issues of network topology and wiring options, both of which are discussed later in this chapter.

Downstream Demultiplexing

Packets received from the Access Network should be demultiplexed so that a particular packet goes to its target application. Without demultiplexing, all packets received from the Access

Network are sent to all devices on the Home Network. This creates problems of excess bandwidth use in the home. Consider the case shown in Figure 7–2.

Figure 7–2
Downstream
demultiplexing problem.

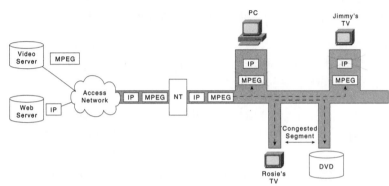

Two packets arrive from the Access Network, an MPEG packet going to Jimmy's TV and an IP packet going to the PC. Packet demultiplexing removes the need for broadcasting every packet on the Home Network and thereby ensures that single copies of packets go only to relevant devices. Otherwise, the TV receives IP packets (which it can't use) and the PC receives MPEG packets. In both cases, the set-top unit must discard packets not intended for it. As a further complication, if Rosie's TV receives traffic from the DVD, then a congested segment occurs between Rosie's TV and the DVD. That is, Rosie's TV is congested with unnecessary IP packets going to Jimmy's TV.

Upstream Multiplexing

In the upstream path, packets from Rosie and Jimmy might need to be forwarded to the Access Network simultaneously. In this case, multiplexing is required. The multiplexing can be as simple as time-division or frequency-division multiplexing. If the offered bandwidth from Rosie and Jimmy exceeds the return-path bandwidth, then some form of contention resolution is required. This gives rise to the requirement for a Media Access Control (MAC) protocol in the home.

Intrahome Networking

If there is substantial intrahome networking that requires quality of service guarantees (for example, DVD to TV), then it will compete with traffic entering the home from the Access Network. If the Access Network contributes relatively little bandwidth to the Home Network, then bandwidth guarantees for in-home traffic are not adversely affected. But if the Access Network and Home Network have nearly the same bandwidth, then there is contention for Home Network bandwidth, which again gives rise to the requirement for a MAC protocol in the home. An example of an Access Network and Home Network that have nearly the same bandwidth is connecting VDSL to the home and using plastic optical fiber within the home, which supports 51 Mbps.

Another issue is to identify a mechanism by which intrahome calls are established. It, for example, is important to ask the following questions: Precisely what are the semantics of a TV requesting a specific program from the DVD? How does a digital camera route real-time output to a specific PC or TV? Some form of signaling for home use is required.

Access to Multiple Networks

The Home Network should provide access to multiple service providers. Access to multiple video feeds, for example, might exist, such as video on demand (VoD) from different content providers going to different viewers in the home. Connecting to multiple Internet service providers could be required, for example, to link to a corporate IP network and the public Internet for recreational use. The Home Network must be able to direct output from the home to the proper content provider.

The requirement to connect to multiple Access Networks also might be a consideration. Jimmy's TVs, for example, might connect to a cable provider, while Rosie's might connect to DBS. Today, each TV has its own set-top box and is bound to that particular network. It is at least preferable that each TV be given the choice of the Access Network to which it connects.

Distance Requirements

The Home Network should cover all parts of the home. Of course, some homes are bigger than others, and greater distances pose problems of power, attenuation, and installation cost. The ATM Forum Residential Broadband Working Group uses a figure of 50 m as a reasonable diameter for a Home Network. Distances greater than this pose problems for some technologies.

Electromagnetic Compliance

Regulations for safety and interference exist in every country in the world, some more stringent than others. In the United States, CE equipment must comply with FCC Part B and Underwriters Laboratory (UL) regulation. This is a key limit for some technologies. Transmitting high-speed data over electric utility lines, for example, using existing technologies requires that excessive power be applied, in violation of FCC regulations.

Ease of Installation

Because consumers have problems with programming VCRs, they could be in for even more interesting challenges installing RBB equipment. Of course, the easiest solution for consumers would be to avoid installation altogether by using existing wiring. Whether or not existing wiring is up to the job depends on what the job is, that is, how much bandwidth is required over what distance. Some technologies explored in this chapter require rewiring, while others don't. Assuming rewiring is required, ease of consumer installation means limiting or avoiding the use of specialized tools or skills. This could preclude certain forms of cabling, such as fiber optics.

Software-end devices will be essentially data networking equipment. Such equipment poses a difficulty for consumers because it requires network addresses, network management capability, and filters.

Currently, an average of two and a half telephone service calls are required to answer customer installation questions for new consumers of Internet services. If a service call takes 10 minutes and the telephone support labor costs the service provider $1.50 per minute, then the customer support calls cost about $37.50. This is roughly the cost of a two-month subscription for Internet service. Technical support of customers is a large factor in why residential-based Internet service has not been a profitable business to date. Many companies have abandoned the residential side in favor of providing Internet services to corporate users. These companies generally have their own in-house network administrators to address users' questions and problems. Auto-configuration will be key in making RBB services profitable enough to be an attractive business proposition.

HOME NETWORK ARCHITECTURE

The Home Network is wired together using one of a variety of contemplated wiring plans. Some plans use legacy wiring, such as currently installed coaxial cable or phone wire. Some plans use new wiring, which requires professional installation. A wireless solution also might exist. The various wiring plans are discussed later in this chapter's section "Wiring Alternatives."

The architectural elements of the Home Network are shown in Figure 7–3. This schematic errs on the side of inclusivity in that not all the elements shown will be part of every Home Network implementation. Some of the illustrated components could be a null function or could be contained in the Access Network or in the terminal equipment instead of in the Home Network.

Figure 7–3
Home Network schematic.

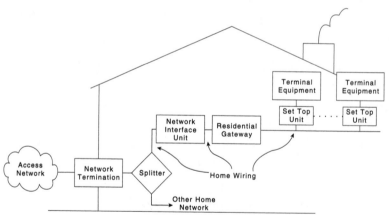

Network Termination (NT)

The *network termination (NT)* is the legal and commercial demarcation point between the carrier and the consumer. This means that if something goes wrong on the carrier side of the NT, the carrier is obliged to fix it. If something goes wrong on the other side, the consumer fixes it, usually by calling the carrier but occasionally by doing it himself. The NT is generally a passive device that performs no changes to the bits or analog signals on the line but can provide some minimal physical-layer functions. In some cases, the NT might not exist as a discrete piece of equipment but is a virtual point at which carrier responsibility ends by agreement.

An example of the NT is the *Network Interface Device (NID)* used by telephone companies. The telco NID is the box on the side of the house to which the telco technician attaches phone wire from the neighborhood telephone pole.

In addition to providing a demarcation point, the NT can perform the functions detailed in the following list:

- **Coupling of home wiring to carrier wiring.** Access wiring terminates here, and Home Network wiring begins.

- **Grounding.** This provides an electrical ground for devices attached to the Home Network.

- **Lightning protection.** This prevents devices on the Home Network from being damaged in the event that lightning hits the Access Network.

- **RF filtering.** In the upstream path, low-pass filters can be installed to prevent unwanted signals generated inside the house from polluting the upstream data path.

- **Splitting.** In the forward path, a single signal can be replicated so that multiple devices in the home can receive the same feed.

- **Interdiction.** As an authentication measure, forward-path signals can be intercepted selectively, such as for premium television channels. When a subscriber pays for a new service, interdiction of a particular program can be lifted.

- **Loopback testing.** To test the integrity of the Access Network, including the drop wire, the NT can provide a test mechanism whereby the carrier injects a signal into the network and receives an echo back from the customer NT. If the echo is successfully received by the carrier, the Access Network is assumed to be functioning properly.

- **Provisioning.** For FTTH systems, the NT might be the device whereby services can be remotely provisioned by the carrier.

Splitter

A second piece of passive equipment in some architectures is the *splitter*. This is a prominent component of ADSL and HFC, and it performs frequency demultiplexing and replication. This device also separates the Home Network into separate pieces, allowing each to use wiring and signaling optimized for a particular function.

In the ADSL case, the POTS splitter peels off the low-speed voice frequencies from the high-speed ADSL frequencies. Its operation is discussed in Chapter 4, "xDSL Access Networks." The result is two sets of internal wiring, one for voice and one for high-speed data. The HFC splitter creates two identical networks; all frequencies are present on both output links. Splitters introduce some frequency attenuation and distortion and are also the source of ingress noise on the upstream. However, splitters perform a useful function to reduce the cost and complexity of distributing signals throughout the home in an inexpensive manner.

Network Interface Unit (NIU)

The NIU is the home device that encodes and decodes bits to and from the Access Network. For HFC networks, the NIU is the cable modem. For ADSL networks, the NIU is the ATU-R.

The following list details the basic functions of the NIU:

- Modulation and demodulation of signals to the Access Network

- Forward error correction

- Enforcement of the MAC protocol

- Regulation of physical-layer–dependent control functions, such as power management in the case of HFC, or rate adaption in the case of ADSL

The NIU pairs with the access node (AN), such as the CMTS (in the case of HFC) or the DSLAM (in the case of ADSL), to enforce the MAC protocol and other control functions. For HFC, for example, the CMTS cooperates with the consumer cable modem to participate in ranging and frequency relocation.

In some cases, it is possible to package the NIU with the NT or with a more intelligent device called the Residential Gateway.

Residential Gateway (RG)

When different networks are joined—in this case the Home Network and the Access Network—a gateway must perform the basic functions of media translation and address translation. One of the architectural options under consideration to perform these functions in the home is called the *Residential Gateway (RG)*.

Media and address translation can also be performed by the NIU but the RG is designed for further software functions. The following list details other potential functions of the RG:

- Downstream demultiplexing

- Upstream multiplexing

- DHCP server

- Network Address Translation

- Address resolution

- IP/MPEG de-encapsulation

- PID remapping

- Authentication and encryption

- Filtering

- Connecting multiple Home Networks

- System management

This list would suggest that the RG is essentially a router. Routing often has proven to be an important scaling technique in networking; it also promotes ease of use. Details of RG functions listed here are discussed later in this chapter, in the section "Residential Gateway Functions and Attributes." In addition, other functions might be provided by a gateway. These functions are discussed in Chapter 8, "Evolving to RBB: Integrating and Optimizing Networks."

Set-Top Unit

After data passes through the NIU or RG, it passes to the *set-top unit (STU)*. The STU is a device that interfaces legacy audio visual (A/V) equipment to the broadband network. A common

STU is the digital TV set-top decoder, in which the STU accepts MPEG packets from the Home Network and translates the packets for presentation on an analog TV set. If a digital TV exists, then the STU can perform other functions, such as connecting a digital camera to the Home Network or controlling an infrared remote control.

A key architectural goal is for the interface between the Home Network and the STU to be standardized to achieve cost reductions. DAVIC specifies this as the A0 interface. With a standardized A0, multiple STUs are possible in the home, each of which can service a different piece of terminal equipment.

Terminal Equipment

The *terminal equipment (TE)* is the generic term used to refer to televisions, personal computers, digital cameras, audio systems. and storage systems. A single Home Network has multiple pieces of TE connected to the Home Network infrastructure, which in turn suggests multiple STUs, one for each TE.

Having multiple TEs and STUs necessitates an addressing scheme within the home, as well as methods to multiplex and demultiplex traffic. TEs might be legacy devices, such as analog TV sets or new devices, specifically designed for digital home use with embedded STUs. The interface between the STU and the TE is dependent on the TE, so there is little standardization work for this interface.

AT ISSUE

Copy Protection

The new generation of audio/visual digital terminal equipment (such as digital video cassette records and DVDs) will provide high-quality displays

continues

of sight and sound. Furthermore, with digital technology it is possible to make copies of content indefinitely without loss of quality. That is, one can make copies of copies, and the last copy will look and sound as good as the original.

This concerns content developers, namely film studios and the music recording industry. An informal, broadbased group called the *Copy Protection Working Group (CPTWG)* is addressing the issue of copy protection. The group is comprised of the major motion picture studios, CE firms, computer companies, and the musical recording industry. CPTWG is investigating the following three aspects of copy protection:

- Device authentication, by which physical devices such as DVDs, digital TV sets, and VCRs are certified as compliant with copy protection. A compliant device will send content only to another compliant device.

- Content encryption, which provides for the encryption of content on the digital medium using techniques that are permitted for export by the United States and other governments. Most governments, including the United States, have rules forbidding the export of high-grade encryption technology. CPTWG encryption must be exportable because of the worldwide trade in consumer electronics (mostly from Asia) and content (mostly from the United States and Europe).

- Copy control information (CCI) includes rules that control the number of copies of a particular piece of content that can be made by a specific device. The content, for example, would specify that zero, one, or an indefinite number of copies can be made by a particular device.

Absent stringent copy protection, content providers believe they cannot enforce their copyrights and will be reluctant to provide their best content in digital form.

RESIDENTIAL GATEWAY FUNCTIONS AND ATTRIBUTES

Given its importance in the Home Network architecture and the many roles it might play, the RG merits closer consideration at this point. This section covers its functions and attributes in more detail.

RG Functions

The need for upstream and downstream multiplexing in the Home Network (both of which would be provided by the RG) has already been discussed. Other specific functions are discussed in following subsections. Keep in mind that these represent a superset of potential RG functions—in different network configurations, some of these tasks could be unnecessary or performed by other components.

Media Translation

In some cases, the physical wiring of the Access Network is identical with that of the internal network. Cable TV networks, for example, use coaxial cable on the drop wire and inside the home. Media translation is not needed in such cases.

However, in most cases the Access Network physical medium differs from the Home Network medium. The Access Network might be wireless, for example, as with LMDS or DBS, but internal distribution might be handled over a wired infrastructure. Similarly, new forms of inside wiring, such as a home's AC power circuits, might be used. Whenever the inside and outside wiring differ, they likely could use different frequencies, have different modulation schemes, and require retransmission. The dissimilarity of the networks forces the need for media translation at the RG, where the networks couple.

Speed Matching

The coupling process requires other physical-layer functions. The speeds of the differing media, for example, typically vary widely. The Home Network might be faster than the Access Network, or vice versa. In this case, buffering is required so that the higher-speed network does not overwhelm the lower-speed network. The RG performs the buffering.

IP Address Acquisition

TE accessing the Internet requires IP addresses so that packets originating on the Internet or corporate networks can unambiguously address the home TE. Readers familiar with the IP protocol suite will recognize that several options exist. Among these are DHCP Server, DHCP Relay, and Network Address Translation. It remains to be seen which of these functions will be performed by the RG.

For readers less familiar with these protocols, the rest of this section provides a brief overview.

Packets from the Access Network do not say "Jimmy's TV" or "Rosie's PC" in the packet header. Some form of globally significant addressing is required, such as an IP address or an ATM address. A mechanism is required to associate the IP address in the packet header with the home device.

What IP address is to be used by the consumer, and by what mechanism is it to be applied to the PC and advertised to the rest of the world? The consumer can manually configure his own IP address, but this is error-prone and cumbersome.

A more user-friendly way to enable address acquisition for the home TE is to have the RG operate the Dynamic Host Configuration Protocol (DHCP) server function. With the RG acting as a DHCP Server, it can supply IP addresses to multiple home computers from a pool of addresses provided to it by the carrier. These addresses can be advertised by the gateway throughout the Access Network and to the Internet at large.

Having the RG perform DHCP Server functions imposes some processing and storage requirements that might not be desirable. Another possibility is to perform DHCP Relay. In this technique, the TE broadcasts a request for an IP address. The RG recognizes DHCP requests and relays them to a server in the Access Network or content provider network specifically addressed in the RG to provide IP addresses. The DHCP Server returns the address to the RG, which in turn relays it to the TE. This centralizes the role of dynamic address assignment in the network.

Another way to provide addressing is for the carrier or RG to identify the end device by a proprietary addressing scheme. A proprietary addressing scheme uses addresses that are not globally unique but that are arbitrarily assigned from the carrier. The proprietary addressing scheme, for example, could use a customer identifier or a billing number.

However, a customer identifier won't do for accessing the Internet because another user might have the same identifier. Therefore, a translation of the proprietary carrier address to a globally significant address must take place. This function is called Network Address Translation (NAT). If this function is required, a logical place for it will be in the RG.

Address Resolution

After an IP address is assigned to the TE, it is stored or cached somewhere so that a quick table lookup can determine the identifier of the TE. In many cases, the job of address resolution can be performed in the Access Network. This could be impractical, however, if the home contains many devices or if the Home Network has an addressing plan unfamiliar to the Access Network provider. In such cases, the job of address resolution could become the function of the RG.

Readers familiar with Ethernet understand the Address Resolution Protocol (ARP) process. IEEE 1394 and possibly other in-home protocols will have their own local identifiers that must be bound to IP addresses. Therefore, the ARP process must be modified to handle these new networks. Address resolution for IEEE 1394 will be a function of the RG.

MPEG and IP Protocol Coexistence

The emergence of digital TV raises the likelihood that both MPEG and IP traffic will be networked to the home. MPEG will convey digital TV, and IP will convey web pages. In some cases, vendors are considering encapsulating one protocol inside the other. It, for example, is possible to carry IP packets inside MPEG frames, and vice versa. Another way to convey both protocols over the same wide area network (WAN) is to use some form of time-division or frequency-division multiplexing.

Whether encapsulation or multiplexing is performed, the multi-protocol case raises the possibility that the RG will be responsible for protocol handling. In the case that one protocol is encapsulated inside the other, for example, a device that de-encapsulates the inside protocol is required. If the protocols are multiplexed, then demultiplexing is required.

A special case of the protocol-handling problem is posed by MPEG. Chapter 2, "Technical Foundations of Residential Broadband," touched on the issue of MPEG PID (program identifier) remapping. MPEG Transport Streams from a single carrier will have unambiguous PID identifiers. However, if a subscriber were to receive digital video from different carriers—such as from HFC and over-the-air broadcast—then the PID numbers might not be unique.

This ambiguity must be resolved before decoding takes place. The RG is a logical place for removing this ambiguity. Means to do so are under consideration by product developers.

Filtering

The gateway can perform a firewall function to prevent unauthorized packets from entering or leaving the home. The packet filters might be controlled by the consumer or the carrier. The consumer, for example, might want to restrict the flow of adult material into the home. Similarly, the carrier might want to prevent the flow of premium programming to customers who don't pay for it.

In the return path, restrictions should be placed on certain packets leaving the home. If, for example, a PC sends data to a printer, there is no reason for that traffic to reach the carrier. Similarly, video traffic from a DVD should not enter the Access Network.

It is also possible to perform filtering in the access node (such as DSLAM or CMTS). By filtering packets in the network rather than the RG, costs associated with filtering (such as memory and processing) are reduced, which is an important consideration for consumer-purchased equipment. However, distributing functions such as filtering into the RGs generally makes the functions more scalable.

Connecting Multiple Home Networks

At least during a transitional phase and possibly over a longer term, multiple Home Networks could exist in a given household. Today, for example, users have phone and cable networks in the home. If a device attached to the phone wire Home Network wants to exchange information with a device attached to the cable Home Network, an RG can serve to connect these separate networks. A household might have other types of Home Networks (such as home electrical wiring) as well making an internetworking device (an RG) that provides media translation and speed matching services even more imperative.

Authentication and Encryption

Carriers might require customers to authenticate themselves to receive service. Encryption would be required for carriers and consumers who want to keep their transmissions private. Both authentication and encryption require some form of intelligent device that can apply relevant protocols, such as Radius or Tacacs, and manage passwords and keys. The RG is a logical place for both processes.

One open issue is whether to authenticate the residence or the device. Authentication to the residence would require only one password for the entire home, whereas device authentication requires a password for each device. It is unclear whether the additional complexity is worth the additional security.

System Management

The RG is a relatively complex piece of consumer equipment, more complex in terms of software than any device currently in the home. For this reason, it is unreasonable to ask the consumer to assume the role of administrator for the gateway. This role is likely to be assumed by the service provider or content provider. The following list details some system-management functions:

- Remote restart and diagnostics capability

- Real-time viewer request processing

- Alarm monitoring and surveillance

- Fault management and recovery

- Software updates and fixes

- Performance management

- Accounting

To accomplish these functions, the RG requires communication with the Access Network or the service provider's management system.

RG Attributes

Based on its likely functions, some of the RG's necessary characteristics begin to emerge.

Operating Systems (OS)

The RG is a computer and therefore requires an operating system (OS) to regulate resource allocation such as memory and bus utilization. Because computer memory is responsible for a relatively large part of the cost of a computer, and because the market for consumer electronics is highly cost-sensitive, the OS is required to be relatively small. In industry terms, it should have a small footprint; that is, the OS should occupy the smallest amount of memory practical. The OS also should accommodate real-time operation and multitasking.

AT ISSUE

Challenge to Microsoft?

Because Microsoft's Windows 95 requires a considerable amount of memory and does not support multitasking, it is not viewed as appropriate for RG use. This has given rise to new developments in the OS and middleware product development space. Many vendors see the RG as an opportunity to compete against Microsoft's Windows in an emerging OS space.

continues

New OS and middleware offerings targeted at digital set tops and consumer electronics include OpenTV™ by Thomson Sun Interactive (TSI)
(http://www.opentv.com), David by Microware (http://www.microware.com),
and PowerTV™ by PowerTV (http://www.powertv.com). Basic OS kernels
are offered, such as VxWorks™ by Wind River (http://www.wrs.com) OS/9™
by Microware (http://www.microware.com), pSOS by ISI, and many others.
Microsoft has responded with a small footprint offering of its own, called
Windows CE.

It remains to be seen how the developments of new OSs that provide
real-time operations and multitasking on a small footprint evolve. Specifically, it will be interesting to watch the impact, if any, this market will
have on the pervasiveness of the Windows/Intel (Wintel) architecture.

Carrier Management

It is likely that the RG will be under some form of carrier management, at least during early product introduction. This means
that the Access Network provider will remotely control the RG.
Carriers would be responsible for software downloads, extract-
ing invoicing information, and customer configuration. Carrier
management is motivated by the complexity of the device and the
desire for user friendliness. The RG thereby would fall under the
model the cable operators use for the set-top box. The box is
owned and maintained by the carrier and is rented to the subscriber. This is different from the consumer electronics model, in
which the item is owned and maintained by the consumer.

Standardization

For the RG to be as low-cost as possible, many parts of it require
standardization. This particularly applies to the Access Network
and Home Network interfaces. Standardization includes modulation components, forward error correction, and MAC protocols. Without standardization, development costs cannot be
amortized over the largest population of boxes.

Bundling with Other Functions

Multiple opportunities exist to bundle the functions of the NIU into the RG. Other options include bundling some functions of the set-top unit. Bundling tends to reduce aggregate costs but reduces flexibility, which has risks during early product rollout.

TOPOLOGY ALTERNATIVES

In addition to network requirements (discussed earlier), network alternatives, including topology alternatives, must be weighed. Some of the proposed Home Networks operate as a bus, some in a star configuration, and some daisy-chained. These configurations are depicted in Figure 7–4.

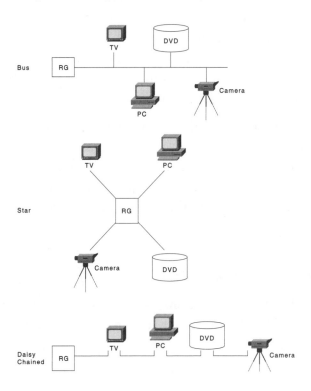

Figure 7–4
Home wiring topologies.

The bus configuration provides economic use of wiring, requires only a single network connection per device, and is well-suited to the broadcasting case. On the other hand, this configuration requires bandwidth arbitration, and demultiplexing is advised. Furthermore, because the bus is a shared infrastructure, it is susceptible to impairments caused by malfunctioning equipment. If, for example, the PC begins to spew extraneous signals, the TV could be adversely affected.

The star configuration isolates traffic per device and therefore can guarantee bit rate per device. This is good for video and for problem isolation. Furthermore, a malfunctioning device would not adversely affect other terminal equipment. But on some occasions, such as broadcast video, broadcast is useful. In this case, packets must be replicated, which creates a processing burden and necessitates the greatest amount of wiring. Finally, a star configuration requires the use of a switch so that packets can exit onto the correct interface. This increases total system cost and introduces an active single point of failure.

Unlike the bus and star configurations, daisy chaining is not common in the commercial LAN world except as a means to connect peripherals to a single computer. This configuration involves the minimum amount of wiring but requires two ports per device. Daisy chaining is similar to the bus but does not require preprovisioning.

WIRING ALTERNATIVES

In addition to topology options, wiring options must be considered in the home. The following list details media under consideration:

- Twisted-pair wires

- Coaxial cable

- Powerline service

- LAN technologies (such as Ethernet or Fast Ethernet)

- IEEE 1394 (also known as Firewire)

- Plastic optical Fiber (PoF)

- Wireless

This discussion categorizes wiring options into three groups: wiring that exists in homes today, wiring that exists in businesses today and is used in LANs but not homes, and new plans that include wireless.

Existing Home Wiring

As noted previously, telephone companies, cable companies, electric utilities, and providers of home automation and security services have for years installed wiring in homes for the transmission of voice, analog television service, and electric power. The following sections explore whether these legacy wiring systems are usable for RBB.

Phone Wire

Telephone customers have twisted-pair wires in their homes. In the United States before 1984, this wiring was owned by AT&T and maintained by the phone company. Now, this wiring belongs to the homeowner. As such, the wire can be used for purposes other than phone service, such as high-speed networking.

Phone wire generally can be considered as a bus topology. The condition of in-home twisted-pair wiring is such that this system typically is not suited for high-speed use. Particular problems are bridged taps and improper splitting, the usual problems found with coaxial cables. Because of impairments, there is little consideration of using this plan for high-speed service. In fact, as noted in Chapter 4, the carrier routinely installs new phone wiring for most xDSL trials. Without new wiring, concerns exist regarding customer satisfaction, especially during a trial.

Coaxial Cable

Coaxial cable in the home is also owned by the homeowner. The physical problems of coaxial cables are similar to those of twisted-pair wires, but home cable is generally newer and often can be used for high-speed networking without substantial upgrading. If, however, the consumer has installed a splitter or an unauthorized set-top box, the chance exists that the wiring is sufficiently disturbed to render it unusable for data.

A final problem with both twisted-pair wiring obtained from the phone company and coaxial cable obtained from the cable operator is that they don't go to all rooms in the house. If you want to leverage your in-home coax for telephone use, you might need to string extra cabling to the kitchen, where you have a telephone hand set on twisted-pair wires but no cable TV. In this case, you would need to rewire your kitchen.

Powerline Service

For ubiquitous service in the home, nothing surpasses electrical lines, also called powerline service. Multiple designs exist, but they all involve the use of an adapter or *powerline modem* on the TE that plugs into the wall outlet of the electrical service.

Electrical outlets are ubiquitous in the home and do not require rewiring. Powerline service is also obviously good for metering applications. Because of the advantages of ubiquity, powerline has been pursued as a home networking solution for years.

A solution has yet to come to market for a variety of reasons, including bit rate, safety, and sharing capabilities. Electrical power lines in the home are unfriendly to high-frequency signals because various electrical components (such as coupling capacitors, inductors, and transformers) act as filters that reduce bit rate.

Power fluctuations induced by attached electrical motors and ingress noise are major problems. Because of such household interference, it is difficult for powerline service to support RBB bit rates and still be in compliance with safety rules as mandated by the FCC and the Underwriters Lab (UL). No powerline solution so far has come to market in compliance for bit rates in excess of a 100 Kbps.

Finally, as shown in Figure 7–5, electrical power lines would be a shared medium among residences connected to a common transformer. If someone were to invent a multimegabit powerline device, it would be shared among 4–10 homes. This creates bandwidth contention and security issues, which must be addressed by an intelligent home device. Nonetheless, because of the simplicity of powerline, it represents an area of considerable product development. One can expect products using powerlines to surface in the coming few years, at least for low-speed applications.

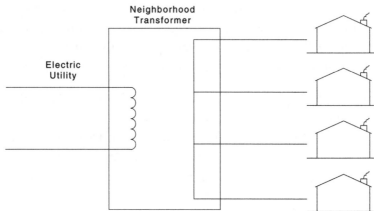

Figure 7–5
Power distribution model.

For now, existing cable, phone wire, and electrical power lines generally are not up to the job of high-speed networking. They have bridged taps, poor insulation, possibly corroded components, and positioning that might not be in the correct place. New alternatives to existing legacy wiring are under consideration.

Local Area Network (LAN) Technologies

Those familiar with internetworking recognize that home networking presents roughly the same problem as local-area networking for business and commercial use. A natural consideration would be to install some form of commercially implemented local area network (LAN) technology, such as Ethernet, Fast Ethernet, Gigabit Ethernet, or ATM. Other possibilities include Fiber Channel, Token Ring, and Fiber Distributed Data Interface (FDDI). These sections examine whether LAN technology can be used for home use.

Home networking differs in several ways from the LAN model. First is the requirement to accommodate broadcast digital TV. No LAN protocol is designed specifically for this role. It is widely believed that video services require isochronous networking. *Isochronous networking* provides bits at a fixed rate over time. Ethernet and its higher-speed brethren do not explicitly provide for isochronous service. ATM is being considered for point-to-point video, but no economic architecture exists for the broadcast case.

Second, the scale of the Home Network is not as large as the corporate LAN. LAN technologies such as FDDI that have isochronous support or high-speed LAN technologies such as Fast Ethernet or Gigabit Ethernet are too expensive for home use.

Third, desire exists to reuse existing legacy Home Networks, such as coaxial cable and phone wiring, rather than use cabling expressly designed for LANs, such as Category 5 UTP.

Most important is the introduction of new participants, such as the CE industry, that do not have a role in the commercial LAN market but see a role for themselves in the Home Network market. The CE industry is endorsing another protocol, called Firewire, that is being standardized as IEEE 1394.

New Wiring and Wireless Proposals

Various proposals for new wiring infrastructures in the home are under study. New wiring promises more speed, more reliability, and lower life-cycle costs at the expense of an upfront commitment to new installation. Chief among the new options are IEEE 1394, also called Firewire, Plastic optical Fiber (PoF), and wireless networks.

Because of this book's view that digital TV is a significant market driver for RBB, this section begins with an in-depth look at the home networking protocol most suited for it, IEEE 1394 and Firewire.

IEEE 1394

IEEE 1394 is a new form of cabling and software architecture for consumer electronics and personal computing. It originally was developed by Apple under the name of Firewire as an interface for multimedia peripherals to Apple PowerPCs. After its submission to IEEE for standardization, the product was given the designation of IEEE 1394, but it often still is referred to as Firewire. Apple and Thomson continue to hold the intellectual property rights, which apply primarily to firms devising integrated circuits to implement the protocol and not to downstream manufacturers of adapter cards and end users.

Firewire has garnered the support of CE manufacturers as a high-speed bus for the attachment of consumer entertainment equipment such as stereo sound equipment, digital cameras, digital video discs (DVD), and digital television sets. Firewire supports the speeds necessary to accommodate multiple streams of digital video and is the medium of choice for CE vendors. Additionally, PC NICs will be available.

The 1394 Trade Association has more than 100 members, including such heavy hitters as Intel, Microsoft, Motorola, IBM, Sun, Philips, Thomson, and all the major Korean and Japanese CE vendors. Apple is the holder of the original intellectual property and intends to ship Firewire on Powerbook and desktop computers in 1998. Sony currently is shipping a digital camera with a Firewire port. Japanese manufacturers such as Sony and Toshiba have announced intentions to ship DVDs and digital decoders with Firewire by Christmas 1997. There is considerable momentum from the CE industry behind Firewire.

Firewire is a high-speed, daisy-chained Home Network with specific software capabilities for autoinstallation. Figure 7–6 illustrates a Firewire network. In this figure, traffic can move between the PC and the DVD, for example, to show DVD images on the PC monitor. A connection is established between the PC and DVD, and traffic flows between the two devices transparently, even though a color printer and a digital audio box are between them. The job of connection establishment is done by the root node, a required element of the architecture that has a global view of all that happens in the Home Network. The root node can be any device in the Home Network; in this case, it is embedded in the home gateway.

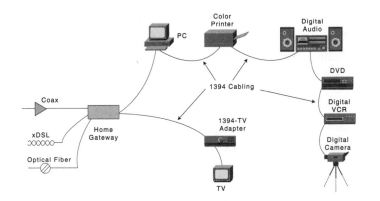

Figure 7–6

Firewire schematic.

Courtesy of Mitsubishi Electric (http://www.mitsubishi.com)

Most devices in the Home Network have two ports to support the daisy chain. The PC, for example, has a wired connection to the home gateway and another to the color printer. Some devices, such as the digital camera, have only one port and therefore must be at the end of the daisy chain.

Note that it is possible to have multiple daisy chains from the root node and to have traffic between them. In Figure 7–6, a second daisy chain from the home gateway includes the TV. So, for example, the DVD also can display images on the TV. Even though the Home Network physically is composed of multiple daisy chains, it logically can be viewed as a single bus.

Firewire is intended in part to simplify home wiring for consumer electronics. Instead of the spaghetti array of wires in the back of stereo sets, VCRs, and TVs, a single cable is used to connect devices in a bus, star, or daisy-chain topology. Other benefits include better pictures and audio as a result of HDTV.

Currently, Firewire specifies a specific type of cabling—referred to here as Firewire cabling—but work is underway to adapt Firewire to other physical infrastructures, such as twisted-pair wiring, PoF, and single-mode fiber.

The following list details the attributes of Firewire:

- Speeds of 98.304 Mbps, 196.608 Mbps, and 393.216 Mbps, commonly referred to as 100 Mbps, 200 Mbps, and 400 Mbps. Work is underway to define an 800 Mbps capability.

- Single-wire cabling for all devices, such as no more separate video cables for audio and video.

- Combined isochronous and asynchronous modes.

- Hot pluggable devices, which are devices that can be added to or removed from the network without interfering with working devices.

- Optional power-passing capability from an IEEE 1394 hub to various pieces of terminal equipment.

- Plug-and-play capability for PC and TV connection.

- Strong support among TE manufacturers.

- Support for digital TV and IP protocols.

Firewire Architecture

Figure 7–7 shows the elements of Firewire's RG. The home-use IP router function defines the address translation of IP to and from Firewire, which has its own numbering plan. The MPEG-2 stream-handling function is concerned with MPEG interaction with Firewire. In particular, this function maps MPEG programs to Firewire channels.

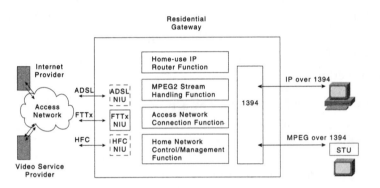

Figure 7–7
Firewire Residential Gateway.
Courtesy of Mitsubishi Electric (http://www.mitsubishi.com)

The Access Network connection function interacts with the Access Network for functions such as autoconfiguration, packet filtering, and authentication.

The control/management function is concerned with connections to carrier management systems, such as fault monitoring, invoicing, and software updates.

This particular schematic envisions modular interfaces to the Access Network. Plug-in cards, such as PC cards, can be used to alter connectivity features. Such modularity is costly, so some other implementations might have a fixed configuration.

DAVIC and the ITU have defined standards for the software adaptation of IP and MPEG over Firewire. Further standardization work for IP is in process by the Internet Engineering Task Force (IETF).

Firewire Principles of Operation

A Firewire network supports daisy-chained and star topologies. Each Home Network consists of a root node, which contains a global view of all devices attached in the home and discerns the topology from control traffic crossing the network. The root node can be any Firewire device, but an RG is a logical place for this function.

The software standard specifies a protocol stack encompassing addressing, timing, bandwidth reservation, and Media Access Control.

Each Firewire-compliant device contains a configuration *Read Only Memory (ROM)* that is embedded in digital set tops, NICs, and consumer electronic devices. This ROM contains the equivalent of an Ethernet MAC address, the speed requirement of the device, and information on whether it operates in asynchronous or synchronous mode. It is the configuration ROM that provides the plug-and-play capability of Firewire. Enough intelligence exists in this ROM to enable it to participate in the configuration process. The result is that the consumer plugs in the Firewire connector and the device is fully capable of participating in the local network.

When devices power on to the network, an event called the *bus reset* occurs. In the bus reset, each device broadcasts the contents of its configuration ROM. The root node listens to all the traffic, distills it, and develops a topology. After the bus reset, the root has a global picture of who is attached to the network, their speed requirements, their requirements for isochronous support, and their identifier.

This protocol has both asynchronous and synchronous support. This is achieved by using a slotted protocol reminiscent of some HFC media-control protocols.

Figure 7–8 shows the frame format for Firewire. Every 125 microseconds, the root issues a frame consisting of time slots for isochronous access and a slot for contention-based access, or asynchronous access. At least 20 percent of the frame (25 microseconds) is saved for terminals wanting asynchronous access. The remainder of the time slot is divided into time slots for use by isochronous users. Terminal equipment desiring an isochronous channel bids for time slots, which subsequently are granted or denied by the root node.

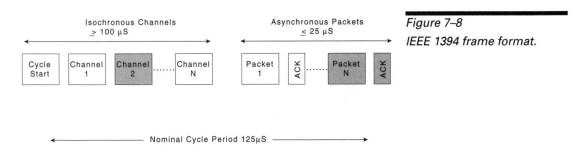

Figure 7–8
IEEE 1394 frame format.

Individual stations send and receive on virtual channels labeled with an 8b channel identifier. Up to 63 isochronous devices can be attached. The channel identifier is assigned by the root and is used by the TE until the next bus reset.

An important function of the RG in the Firewire model is to bind the 1394 channel identifier to an MPEG PID or IP address.

Firewire Issues

Firewire represents a very high-end Home Network architecture that is particularly well-suited for video transmission, including broadcast TV. Nonetheless, despite the plug-and-play feature, simplified cabling, speed, and support of the CE industry, Firewire faces hurdles.

The following list details some of the challenges for Firewire:

- Range is limited to 5 m between devices; repeaters are required beyond that. An updated standard, called IEEE 1394b, will specify Firewire over optical and plastic fiber. This will solve the reach problem, but the standard is not scheduled for completion until the middle of 1998.

- Rewiring in the home is required, or the Firewire cable will be left dangling, as is currently the case for CE and PC wiring.

- Firewall requires some form of RG. The root device is an inherent part of the architecture.

- The cost is significant compared to legacy systems. The Firewire cables are new, the interfaces are new, and there is a requirement for new devices such as set-top units, the root processor, and home rewiring.

Further information can be obtained from the 1394 Trade Association web page at `http://www.Firewire.org`.

VIEWPOINT

1394b Versus Fast Ethernet

The current deadline for an IEEE 1394b specification—which will provide for a long-distance Firewire—is now set for the middle of 1998. This appears to give an opening to the application of LAN technologies (particularly Fast Ethernet) in the home. Fast Ethernet is available on PCs today, and its software issues are well-known. Video services over the Internet can be provided quickly, for example, by encapsulating MPEG packets within IP packets.

If the market for RBB can grow sufficiently just on Internet access, then it is possible that the LAN technologies can achieve a substantial foothold in the home and become de facto standards.

Plastic Optical Fiber (PoF)

Another form of new cabling under consideration is *Plastic optical Fiber (PoF)*. Instead of using a metallic infrastructure in the home, proponents of PoF and FTTH propose the use of a clear plastic that can transport optical signals in the home at a bit rate of 51 Mbps. The medium is very lightweight, immune to in-home interference, and very inexpensive. PoF is one of the home-wiring options under consideration in the ATM Forum Residential Broadband Working Group, primarily because of its immunity to home-based RF interference.

Advocates of PoF propose to use a light-emitting diode (LED) operating at 50 Mbps. The PoF interface is intended for link lengths of up to 50 m.

One problem with plastic is that it is brittle. Copper is ductile, and it bends. For PoF, a minimum bend radius stands at 25 mm and a maximum number of turns. PoF can withstand 20 quarter turns (90° each turn) before attenuation exceeds 9 dB loss. Unfortunately, in many homes, there will be a requirement to bend the medium more than 20 times as it winds its way through the baseboards, ceilings, and walls. For this reason, PoF might be usable only in a star topology. Fatigue is also a factor in bending glass or fiber. Small fissures form over time, which increases attenuation.

The following list summarizes the features of PoF:

- Speed up to 51 Mbps

- Distances to 50 m

- Immunity to household RF interference

- Plug-and-play capability for PC and TV connection

This list details some of the major issues of PoF:

- Little support among TE manufacturers

- Lack of definition for software elements, such as Media Access Control, timing, and management

- Limited bend radius and number of bends

- Speed limited to 50 Mbps

The problem with 1394 and PoF is that they require new wiring. This negatively impacts their appeal, but they provide better service. Rewiring is discussed later in this chapter in the section "Commercial Issues."

Wireless LAN

If home rewiring is too much of a problem, the wireless option could be an alternative. Wireless has obvious advantages of mobility and ease of installation because no cables must be installed. Wireless LANs are used in office buildings to connect printers, PCs, and routers. These LANs can penetrate walls and ceilings with 10 Mb of service. The same technology can be used for home use, provided that costs can reach consumer electronics levels.

But wireless has been a difficult problem for LANs and so far has enjoyed relatively little penetration, even for commercial use. The following list details the major reasons for this lack of penetration:

- Interference and noise occur due to noncompliant 802.11 devices occupying the same bandwidth, such as microwave ovens and other unlicensed spectrum users.

- Power management is necessary to carefully regulate the output of wireless transmitters so they don't overwhelm other legitimate users. This is especially difficult when devices are in motion.

- Because of multiple international regulations, spectrum use differs among countries and limits marketability of some products. Differing regulations on safety also have an effect.

- Cost is considerable.

However, because of the attractiveness of wireless networking, particularly for residences that have a premium on ease of use, wireless LAN technologies are under consideration in the ADSL Forum for home distribution.

Several proprietary approaches to wireless LAN exist. One published option being standardized by IEEE is 802.11.

IEEE 802.11

The IEEE has been in the process of developing standards for wireless LAN networking for years.

The following list details some of the key features:

- A 2 Mb full-duplex service

- Asynchronous and isochronous service

- Use of multiple frequencies and multiple modulation techniques

- Specification of a MAC protocol

IEEE 802.11 endorses three physical-layer specifications for usage in the United States. These specifications are detailed in the following list:

- Frequency-hopping spread spectrum using radio spectrum

- Direct sequence-spread spectrum using radio spectrum

- Diffuse infrared, which does not require an uninterrupted line of sight between the transmitter and the receiver

In the United States spread-spectrum devices for private use operate in various unlicensed spectra. The following list details these ranges:

- 902 MHz–928 MHz

- 2.4 GHz–2.4835 GHz

- 5.725 GHz–5.875 GHz

Different frequencies are used in Japan and Europe.

Wireless Issues

The software issue for wireless networking that the 802.11 standards people had to deal with involves how to create a common MAC protocol for the multiple physical layers that is both spread spectrum and infrared. Another problem is how to deal with roaming devices and battery-powered devices. Roaming refers to the fact that the device can be mobile, one of the greatest benefits of wireless. To conserve power consumption, an important characteristic of roaming devices is that they be able to go into sleep mode whenever they are not receiving or sending equipment. 802.11 has provisions for buffering messages for a sleeping device, waking it up, and forwarding packets to it.

Keep in mind that the Home Network typically does not have a roaming or battery-consumption problem. As a result, although 802.11 addresses these problems, it does so at a cost that is not required for home use.

Another issue with 802.11 is that it is restricted to 2 Mbps, which precludes its use for video. Scientists at Lucent Technologies, however, announced a new technology called *DS/PPM* to obtain 10 Mb. DS/PPM transmits in the 2.4 GHz band and uses Direct Sequence Spread Spectrum. A 10 Mb wireless is state-of-the-art for wireless LAN use in unlicensed spectra. Product rollouts have not yet been announced.

This development symbolizes the eternal optimism surrounding wireless networks. Wireless LAN technologies and their corporate sponsors so far have failed to compete in terms of speed and cost with wired solutions in corporate LANs. But the allure of mobility and ease of use makes it likely that wireless LAN product development will continue and that some forms eventually will become useful in the home-distribution network.

COMMERCIAL ISSUES

Before home networking becomes commonplace, issues apart from technology must be addressed by vendors and carriers.

Partnering

The Home Network is turf on which many industries will converge. Individually, none of them so far seems to have all the req-
uisites to succeed. Table 7–1 indicates some industries and their
strengths and weaknesses with respect to home networking.

Table 7–1   *Candidate Home Networking Industries*

Industry	Companies	Strengths	Weaknesses
Consumer electronics	Mitsubishi, Sony, Sega, Nintendo, Toshiba, and Panasonic	Consumer channels Branding Capability to leverage content sales	Lack of Internet expertise Lack of carrier experience
Digital TV set-top manufacturers	Thomson, General Instruments, and Scientific Atlanta	Strong MPEG Alliances with carriers	Lack of Internet expertise
Internet vendors	Cisco, Bay, 3Com, and Ascend	Internet expertise Alliances with carriers Systems experience	Lack of consumer channels and branding Not accustomed to low margins Lack of broadcast video experience
Telephone companies	Local exchange carriers Long-distance carriers	Branding Retail sales Systems experience	Lack of a multichannel infrastructure; digital TV will be tricky
Cable operators	TCI and Time Warner	Consumer marketing entertainment Broadband infrastructure; can do digital TV	Lack of capital Return-path problems

Table 7–1 *Candidate Home Networking Industries, Continued*

Industry	Companies	Strengths	Weaknesses
Wireless operators	DBS companies Telephone companies CellularVision, Heart-land, and ATI	Low rollout costs MPEG digital channel capacity Local content, except DBS Branding, for telcos and DBS	Lack of branding and capital (except for telephone companies) Return-path problems
Broadcasters	NBC, ABC, CBS, PBS, Fox, BBC, and NHK (Japan)	Low rollout costs; owners of free spectrum Branding Local content	Lack of networking experience Return-path problems

The CE industry, for example, has strong consumer channels, understands consumer-support issues (such as how to deal with returns), and implements excellent branding. This industry also has strong computer processing experience with recent success in games. The problem is networking. Products of the CE industry do not require addressing, software management, filtering, and other functions familiar to networking vendors and carriers.

Networking vendors face marketing challenges when dealing with the consumer market. They are accustomed to higher margins on hardware products because they do not sell disposables to subsidize equipment sales. As a contrasting example, video game vendors recover the cost of the game players by selling game cartridges. Networking vendors have few consumer channels, no infrastructure for customer support, and, therefore, little consumer branding.

Given their complementary weaknesses and strengths, as well as the scale of investment in product development and risk, different industries are likely to form new alliances to address the Home

Network problem. The key to winning in the home market is the development of strong partnerships. Exactly which partnerships will be formed and for which markets remains to be seen.

Fixed RG Versus Modular RG

For homes attaching to multiple Access Networks, each Access Network will require a network-specific NT and NIU. A cable modem, for example, cannot be used for xDSL systems. This raises an important packaging question for the RG: How will it accommodate multiple Access Networks? In a related question, if a consumer changes Access Networks, is he required to change the RG, or will the RG have a modular design that can accommodate multiple Access Networks with the addition of a PC card or a similar attachment?

The flexibility of a modular design will increase the cost of the RG. Additionally, a modular RG will do little to reduce churn. Carriers must consider, for strictly commercial reasons, whether they need to retain the account control that a carrier-specific NIU will provide.

Rewiring the Home

Rewiring a home is a big problem for most homeowners. Renters could be precluded entirely. Old homes could have asbestos problems. Proponents of LANs and Firewire must confront arguments against rewiring when constructing their business case.

So far, several responses to these arguments have taken shape. First, more than a million new homes are built in the United States annually, and millions more are remodeled. If a fraction of these were to be rewired by the developer, or if conduits could be installed to enable easy future wiring, some of the problem would be alleviated. Second, the bulk of home entertainment and work-at-home applications likely will be only in one or two rooms of the home. There is no need to have a device in every room, such as the kitchen and bathrooms. If there is a home entertainment center and a work-at-home center, then a high-speed point-to-point network in the home would suffice for most purposes.

Finally, the possibility exists that new wiring can be installed but not through walls. This means wires would be left dangling, taped to baseboards, and strung around doorways. This is how lots of stereo wiring and all PC wiring is done. Some say there will be more of this approach because the value of the new wiring is so great.

In any case, the carrier installing the Access Network probably will be responsible for the Home Network. As such, the carrier likely will perform some form of installation for a fee. Whether or not the carrier's responsibility for installation makes rewiring scenarios untenable from a business perspective remains to be seen.

Full-Service Home Networking

Because separate networks exist for phone, cable, power, and home automation, the question arises whether there is a strong market need to integrate services on a single Home Network architecture, that is, a full-service network.

The perceived advantages of full-service home networking to the customer are simplified ordering, wiring, and maintenance. The disadvantage to the consumer is that a single network makes the consumer vulnerable to a single-network outage and possibly could make the customer captive to a single Access Network provider.

The candidates for full-service Home Networks are coaxial cable, PoF, and Firewire. Only cable would not require home rewiring, but use of it requires subscription to HFC or FTTC Access Networks.

The advocates of full-service networking include the ATM Forum, the HFC industry, and supporters of Fiber to the Home. The ATM Forum sees ATM as the best way to offer latency control to applications. The HFC industry's interest is motivated by the fact that it has the most widespread physical infrastructure to offer both digital TV and data services to the home. Supporters of FTTH contemplate end-to-end service over fiber-optic or PoF cabling.

ADSL and various wireless local loop technologies do not offer the prospect of full-service home networking for both broadcast video and data. This might not be a disadvantage to these industries, because the economic case for a full-service network in the home is not yet established.

STANDARDS ACTIVITY

The subject of home networking is under considerable study in various standards forums. Chief among these are DAVIC, Vesa, ATM Forum, ADSL Forum, IEEE 1394 Trade Association, and Telecommunications Industries Association (TIA).

Digital Audio Visual Council (DAVIC)

DAVIC 1.0 specifies an audio/visual distribution architecture for the Home Network, called the *Service Consumer System (SCS)*.

Figure 7–9 illustrates the key components of SCS. Traffic enters through the NT and connects to a *set-top box (STB)*. The STB consists of two functional entities, the NIU and set-top unit

(STU). The NIU is the modem function discussed previously. The STU is responsible for the value-added software features needed to connect to a piece of TE. These functions include session establishment, security enforcement, MPEG filtering and decoding, and IP processing. After considerable discussion, there is a consensus that the A0 interface will be IEEE 1394.

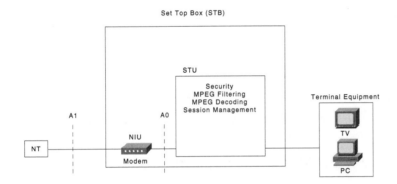

Figure 7–9
DAVIC SCS schematic.

Courtesy of Digital Audio Visual Council (http://www.davic.org)

Another packaging option for the STU function would be to include it within the TV or PC, in which case the NIU is a stand-alone box with an external A0 interface. This permits modularity of the NIU, which means that the consumer can change carriers without changing the STU or TE.

The problem with the model shown in Figure 7–9 is that it does not account for the case of multiple devices in the home—it can accommodate either the PC or TV illustrated, but not both. A successor architecture, depicted in Figure 7–10, includes a new interface called *A1 ** and a new function called the *User Premises Interface (UPI)*. This interface does accommodate multiple terminal devices.

Figure 7–10
*DAVIC SCS multidrop
schematic.*
Courtesy of Digital Audio Visual
Council (http://www.davic.org)

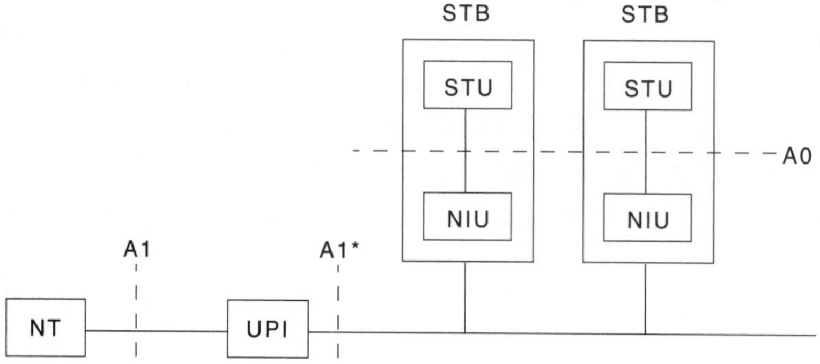

The inclusion of an RG in the Home Network would directly address the role of the UPI. As of this writing, however, DAVIC has not specified details of either the UPI or A1*.

ATM Forum RBB Working Group

The RBB Working Group is one of a number of working groups in the ATM Forum. The intent of the Forum is to specify how ATM is to be implemented in end-to-end architectures. The objective of the RBB Working Group is to specify the interface between the home and various ATM Access Networks, including HFC, xDSL, FTTC, and FTTH, and between devices inside the home. Inside the home, the RBB Working Group is considering PoF using ATM25 or ATM at OC-1 speeds (51 Mbps). The motivation for use of ATM is to provide bit-rate guarantees for video traffic and virtual channels for multiple types of TE in the home.

The fundamental assumption is that ATM is the carrier from the service provider all the way to the TE. This requires the presence of an ATM network in the home and ATM-compatible TE.

ADSL Forum

The ADSL Forum contains a working group that specifically addresses the issues of the Home Network. Figure 7–11 depicts a schematic of the customer premises reference model.

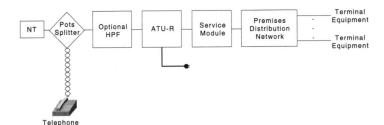

Figure 7–11

ADSL Forum customer premises reference model.

Courtesy of ADSL Forum
(http://www.adsl.com)

Traffic enters the home through the NT. Legacy voice traffic immediately is separated using a splitter. The splitter is a low-pass filter that enables voice traffic and power to pass through to telephones in the home. In the event of power loss in the home, voice service still can operate. Unlike HFC and high-speed wireless Access Networks, xDSL places special emphasis on legacy voice telephony. Therefore, the ADSL Forum has had considerable deliberation on the placement and functionality of the POTS splitter.

One question given extensive thought is where to put the splitter. It can be integrated with the NT, integrated with the ATU-R, or standalone. Each option is acceptable to the forum, but it recommends that the splitter be standalone. The problem with NT integration is simply that it won't fit. The difficulty with ATU-R integration is that the phone might be near the NT. In this case, it could be necessary to backhaul voice service to the NT. Extra wiring also can cause coupling problems, so the current recommendation is that the splitter be standalone and relatively close to the NT.

Traffic above voice band (3,400 Hz) is passed through to the ATU-R, which is a modem concerned with CAP, DMT, and forward error correction. Consideration is being given to whether traffic entering the ATU-R should pass through a high-pass filter (HPF), further ensuring reliability. An HPF at each device would seem expensive but could provide better service by reducing noise into the ATU-R. Another option for the HPF would be to embed it in the splitter. So far, the high-pass filter is not explicitly part of the reference model.

The service module adapts traffic from the ATU-R for use on the premises network. In many cases, this can be a null function. The premises-distribution network is simply wiring and corresponds to the home-distribution network in the ATM Forum case. The premises distribution network is likely to be phone wire, either legacy phone wire or newly installed phone wire.

Because substantial broadcast video use of ADSL is unlikely, other, lower-speed premises-distribution networks can be considered, such as wireless networks.

1394 Trade Association

Supporters of Firewire have organized themselves into a trade association called the 1394 Trade Association. The group's function is the continued technical and marketing support for Firewire and various enhancements.

Current features and standards of Firewire were discussed earlier in this chapter, in the section "IEEE 1394." Some of the enhancements under consideration include the definition of higher-speed Firewire and the use of Firewire over cabling other than the original Firewire cable. These media include Category 5 wiring and single-mode fiber. With single-mode fiber, there is the expectation that Firewire can be used over very long distances, thereby reducing the requirement for ATM and other WAN protocols.

Video Electronics Standards Association (VESA)

VESA is an industry group established in 1989 with more than 290 current members. Until recently, the organization's focus has been on mechanical and electrical characteristics of CE devices. VESA has issued papers covering areas such as display parameters for monitors, timing specifications (so that progressive content can be displayed on interlaced monitors), and the operation of flat panel displays.

Recently, however, the organization has turned its attention to systems problems, in particular the formation of a Home Network group. Given the group's recruitment of CE vendors, this group of VESA will emphasize the systems aspects of the delivery of home video.

SUMMARY

The Home Network is the convergence point where computing, networking, broadcasting, and consumer electronics come together in the largest quantities. Because it is situated on the customer's premises, the Home Network has the special challenge to be as user-friendly as possible. Other market, technical, and regulatory challenges are summarized in Table 7–2.

The combination of digital video and Internet access creates complex issues of multiplexing, demultiplexing, signaling, Media Access Control, address resolution, and other issues that are properly resolved with a router-like device, the RG.

The next chapter examines some end-to-end software issues required to combine Access Networks and Home Networks into a coherent economic system.

Table 7–2 *Central Issues for Home Networking and RBB*

Issue	Challenges	Facilitators
Home Network equipment	Integration of NT, NIU, and RG functions is needed for cost reduction. Modularity is needed to support multiple Access Networks and Home Network topologies and technologies.	Collaboration and competition exists between multiple vendors, including game players, TV decoders, and internetworking vendors.
Use of existing home wiring	Impairments might be excessive. Cost of upgrading and repair, where needed, is high.	A minimum cost scenario exists. Wiring is owned by homeowner, who is responsible for maintenance and upgrades.
Intelligent Residential Gateway	Costs, choice of operating systems, programming complexity, and standardization are troublesome.	RG facilitates scaling, a wider variety of services, ease of use, and carrier maintenance.
Standards	A consensus is needed on a multitude of functions. Coordination is required among numerous standards bodies and geographical areas.	Basic standards are established for Service Consumer System (DAVIC), use of ATM (ATM Forum RBB Working Group), ADSL reference model (ADSL Forum), and Firewire (IEEE).
Support for video	Support requires high-speed networking, quality of service controls, and signaling. (How does TE inform network of speed requirements?)	Support enables funding via the familiar advertising model.
Full-service networking	This approach makes it difficult to put all services on legacy networks. The consumer loses all services when the one big network goes down. This option also doesn't exist now; do we need it?	Full-service networking provides account control for the carrier. This option makes life simple for the consumer. The lowest cost option is available to the consumer.

Table 7–2   *Central Issues for Home Networking and RBB, Continued*

Issue	Challenges	Facilitators
Market	How much responsibility will the carrier accept for inside wiring, and what will be the costs to the consumer?	Faster home wiring will improve services, especially for video. This is necessary to support faster Access Networks.
Industry structure	Multiple industries are involved, each with differing requirements.	Partnerships and alliances are forming.
Business/ Financial	Cost of and responsibility for rewiring are troublesome. Cost of new adapters and RG is high.	Maybe rewiring isn't such a problem, given recent trends in new home construction and home remodeling.

REFERENCES

Articles and Books

ADSL Forum Working Text 011-R2. "Interfaces and System Configuration for ADSL: Customer Premises." December 1996.

Dearnaley, Robert, Steven Hughes, and Alan Quayle. BT Labs, "The residential EMC environment and its implications for in-home transmission systems." *ATM Forum/96-1047R1*.

Dodds, Philip. 1995. *Digital Multimedia Cross-Industry Guide*. Focal Press, ISBN 0-240-80205-5.

Fujimori , Taku H., Richard K. Scheel, Sony Corporation. "An Introduction to an IEEE1394-Based Home Network." *ATM Forum/95-1378*, October 1995.

Moulton, Grant (General Instruments). "Spectral Allocation for the Optimal Delivery of Broadband Signals over In-home Coaxial Cable." *ATM Forum/96-0483rev1*, August 1996.

Quayle, Alan, and Steve Hughes. "The Nature of the Home ATM Network." BT Labs, *ATM Forum/97-0163*, February 10-14 1997, San Diego.

Internet References

http://www.adaptec.com/

http://www.davic.org
(DAVIC is everywhere in the RBB space. Part 5 gives the DAVIC view of the Home Network, which they call Service Consumer System.)

http://www.fireflyinc.com
(The host site for 1394b, the updated version of Firewire.)

http://www.Firewire.org
(The home of the 1394 Trade Association. Mostly cheerleading, but the FAQ and technology are here as well.)

http://info.gte.com/gtel/sponsored/rg/webspec.htm
(GTE's idea of a residential gateway.)

http://www.ti.com/sc/docs/msp/1394/tech.htm
(Technology overviews of Firewire from Adaptec and Texas Instruments, which stand to gain a lot of silicon sales.)

http://www.vesa.org
(The home of the Video Electronics Standards Association.)

CHAPTER 8

Evolving to RBB: Integrating and Optimizing Networks

Access Networks and Home Networks have made tremendous strides over the past few years. The research and development work of companies, standards bodies, and academics worldwide has resulted in the creation of technologies to move data at multimegabit speeds over thin wires and thin air. The solution of discrete technical challenges is progressing at a rate faster than many experts would have predicted a few years ago.

The development of Access and Home Networks is at a high enough plateau that the next set of challenges can be confronted. Among these challenges are networking issues that concern how the various components are integrated to create end-to-end systems. This chapter overviews the networking challenges that arise when individual services and networks come together as an integrated whole to provide RBB service.

In coming together, content providers must consider distribution decisions, and Access Networks must select content. System-design and deployment strategies are considered from both directions—both content and distribution—in this chapter. A case study of a full-service network is offered to illustrate integration challenges within a specific context. The chapter concludes with some tools for forecasting the potential RBB networks that are likely to succeed.

NETWORKING ISSUES

Trials of RBB networks to date primarily are concerned with resolving transmission challenges, such as modulation and forward error correction. Before large-scale deployments are made, however, tough networking issues must be solved. These networking issues include scaling, ease of use, security, and the capability to support new applications. End-to-end cooperation—from the end system through the Access Network to the content provider—is required to implement networking solutions. To service millions of users, merely continuing to throw bandwidth at the problem will not suffice. One can argue that today the Internet faces scaling problems even at current bit rates.

Server Caching

An important scaling technique is *server caching*, which is the replication of content at geographically dispersed places. Instead of millions of users converging on a single server, content of a popular server is replicated and accessed by users elsewhere. Figure 8–1 illustrates the concept of server caching, with the original cache being replicated via the Internet and Access Network. The user accesses the server cache through the Access Network.

The following list details the tasks associated with server caching:

- Ensuring that the original server and the caching server have the same content

- Ensuring that the process of synchronizing the databases doesn't generate excessive traffic

- Ensuring that the consumer reaches the geographically nearest server with current information without having to explicitly identify the server

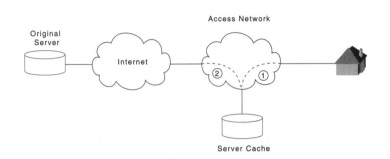

Figure 8–1
Server caching.

A number of approaches can be used with server caching. The first is to have the original source periodically multicast updates to the replicating servers. This requires IP multicast in the network and creates the potential problem of updating servers unnecessarily. Even servers that service relatively few requesters would receive the periodic updates.

A second approach is to have the local router forward requests to the nearest server cache, as illustrated by Path 1 in Figure 8–1. If an updated record exists, it is returned to the user. If no match exists, then the request for information is forwarded to the original server via Path 2 in Figure 8–1, which answers the request. In so doing, the replicating server notices the response and caches it. The cached entry is aged in a time appropriate for the nature of the content, the request rate for the page, and the cache server resources. All requests in the interim are answered by the replicating server until the record is removed. The entry then is refreshed, possibly with a modified record after the next cache miss. This technique enables the replicating server to keep a minimum amount of information, which minimizes backbone congestion.

Addressing Issues

The scale of RBB creates a major problem with address proliferation. Each residence—in fact, each device in the residence—must be identified uniquely.

This requires the management of large address spaces, which in turn necessitates the following question: How are customers to be addressed in a systematic manner, and how are addresses to be supplied to the residence?

In the case of the telephone network and postal services, addresses are supplied by the carrier. You get your phone number from the phone company and your zip code from the U. S. Postal Service. The addressing plans provided by these service providers assist in routing traffic to the residence.

The problem in the case of RBB is that situations exist in which addressing is supplied by the content provider. An IP user at home, for example, must be a part of the addressing domain of the corporate network. In the case of broadcast television, the PIDs are under the control of the content provider. If the Access Network provides access to multiple video content providers, there is the potential problem of ambiguous PIDs, because two or more video providers might use the same PID for different programs.

In IP cases, the internal Home Network is required to have IP addressing. This can be provided by the carrier by supplying globally significant addresses or private addresses. A mechanism is required to provide such addresses to the home, however. If private addresses are provided, a network translation function in the home is required. DHCP is the mechanism most likely to be used to support address acquisition. (DHCP is discussed further in Chapter 7, "In-Home Networks.")

Always Connected Model Versus the Terminal Server Model

Today's dial modem users connect to the Internet through a terminal server. A terminal server answers the telephone call from the modem, authenticates the call, and facilitates a connection.

The terminal server model has served consumers well for online services and the Internet. But for many RBB applications, another model is preferable in which the consumer is always connected to the network, such as a business local area network (LAN). The following list details some of these applications:

- Personal web servers in the home, a trend encouraged by faster speeds, large PC disks, authoring freeware, and second phone lines.

- Push-mode data.

- Internet telephony. The consumer must be connected to the Internet before receiving an Internet phone call.

- Business applications, which are always connected to a local area network (LAN), such as software updates and push-mode data.

When permanently connected, the consumer is relieved of the dialing process, making the system easier and faster to use. Another advantage of always being connected is that the carrier can perform certain functions, such as software modification, without involving the consumer. This facilitates ease of use.

The drawback of always being connected, however, is that an idle consumer device consumes some network resources, a little bit of bandwidth, and a little bit of memory. But on average, RBB networks seem to be headed toward the always connected model because of ease of use and new applications.

Transport Protocol

When a data or video packet is sent to or from a residence, it is encapsulated in a transport protocol. The selection of transport protocol represents a networking challenge because competing approaches are available, each of which has advantages and disadvantages. Systems using different transport protocols cannot interoperate directly. The current candidates are IP, MPEG, and ATM. With IP encapsulation, all packets are sent over the Access Network to the home in IP packets. MPEG packets are encapsulated inside IP. The home obtains the IP packets, determines that the contents are really MPEG, de-encapsulates the packets, and forwards the MPEG packets inside to the set-top decoder function.

With MPEG encapsulation, the reverse process is used. MPEG is sent to the home and decoded by the digital set-top box. IP packets are indicated by a PID number. MPEG demultiplexing then forwards the data PIDs. This approach is used by MCNS cable data consortium, which has specified that IP data packets sent to the residence will be transported in MPEG frames.

With ATM encapsulation, both IP and MPEG packets are transported in ATM cells using AAL5 adaptation. MPEG streams are differentiated from IP packet flows based on virtual channel (VC) numbers. Multiple VCs would exist between the carrier and the residence and would be bundled together in a virtual path (VP). Because each residence has a VP, a standardized signaling channel can be used from each home to the carrier. The binding of ATM VCs to individual IP or MPEG flows can be set manually (as it is during trials) or established using ATM signaling. ATM encapsulation is the choice of DAVIC, the ATM Forum (not surprisingly), and the ADSL Forum.

Microsoft has ATM support in WinSock (its communications interface for Windows) and therefore can perform the de-encapsulation function inside the PC.

The advantages and disadvantages of each encapsulation method are summarized in Table 8–1.

Table 8–1 *Comparison of Encapsulation Methods*

Protocol	Advantages	Disadvantages
IP	Enables native transport over the Internet. Would permit content providers to bypass broadcast TV or cable by distributing content directly over the Internet.	If more video bits than data bits, involves the greatest amount of de-encapsulation among the three alternatives. Has possible issues with IP addressing, which could require upgrading to the next generation of IP (an expensive process). Packet latency is not guaranteed, which places extra buffering and time-synchronization complexity on the MPEG decoder.
MPEG	Nearly all compressed digital video content—satellite, DVD—is already in MPEG TS format. Well-suited to point-to-multipoint access networks. One-way—good for video!	Not amenable to a point-to-point environment, and has no addressing that identifies a specific user. One-way—bad for transactional!

Table 8–1 *Comparison of Encapsulation Methods, Continued*

Protocol	Advantages	Disadvantages
ATM	Permits transparency of multiple protocols. Allows for ATM quality of service. Gives carriers the flexibility to establish pricing based on quality of service.	Has few ATM terminal equipment models today, especially for video. Involves costly signaling and addressing methods. Not well-suited to point-to-multipoint access networks.

Navigation

With 200 television channels and many options for web and video on demand (VoD), consumers face the problem of finding specific programs. TV viewers are familiar with *barker channels*, which are TV channels that list programming for other channels.

In the SDTV case, multiple programs might exist for every 6 MHz channel. Identifying both the frequency and the PIDs within that frequency will be required for viewers to select a program or a movie. This represents a new level of sophistication for electronic program guides.

Tunneling

RBB will afford consumer access to multiple content providers through a common Access Network. As an analogy, you can telephone anyone by using a common phone network. The generic term used in the voice case for this ubiquitous access is *dial tone service*. The same term and concept can be applied to data service.

A parent, for example, might want to be connected to work, while his daughter might want to surf the Internet for recreational or educational use. The father could use an Access Network that provides transit to his corporate network. The corporate network should appear to him just as if he were connected to a LAN at work. This network always would be connected, fast, secure, and accessible to all corporate applications. The daughter, who simply needs Internet access, has no particular service requirements other than connectivity.

The work-at-home requirement involves a number of complications. First, the connection must be authenticated by the corporate network. Second, the work-at-home user should be a part of the corporate IP numbering plan rather than the Access Network numbering plan. Both of these requirements exist for security reasons and do not apply to the casual Internet user. How can a single Access Network interface accommodate both users?

One technique is called the *Layer 2 Tunneling Protocol (L2TP)*. Its role is illustrated in Figure 8–2. The user establishes a PPP connection to an authentication server on the Access Network. The L2TP provides open connectivity to multiple content providers, which the user specifies in the PPP packet. In the voice case, the user specifies the destination by dialing a phone number. In the IP case, the user specifies a destination domain, such as "cisco.com."

Figure 8–2
*Layer 2 Tunneling Protocol
(L2TP).*

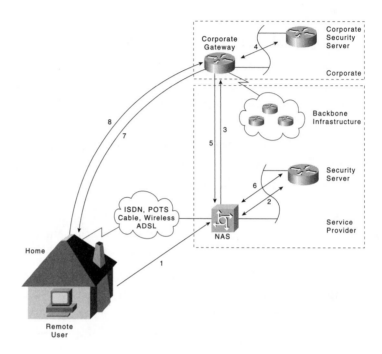

Figure 8–2
Layer 2 Tunneling Protocol (L2TP).

The L2TP process, described in Figure 8–2, follows these steps:

1. The remote user initiates PPP, and the Network Access Server (NAS) accepts the call.
2. NAS authenticates the remote user (by verifying user name, password, and destination domain) using a security server.
3. NAS initiates an L2TP tunnel to the desired corporate network. A tunnel is an encapsulation of each packet within another IP packet.
4. The corporate gateway confirms acceptance of the call and the L2TP tunnel.
5. NAS logs acceptance.
6. The corporate gateway authenticates the remote user and accepts or rejects the tunnel.

7. The corporate gateway exchanges PPP information with the remote user. An IP address is assigned by the corporate gateway to the remote user.

8. End-to-end data is tunneled between the remote user and the corporate gateway. The remote user is logically connected to the corporate internal network.

The daughter's connection can be made directly to the public Internet without an L2TP tunnel. The NAS forwards her packets to the Internet without intervention of the security server. For further details on L2TP see the Cisco home page at http://www.cisco.com and search for "L2TP."

DISTRIBUTION SCENARIOS FOR CONTENT PROVIDERS

In the integration of content and Access Networks, content providers have the option of distributing through the Internet or over broadcast networks. The choice of Access Networks is governed in part by the technical nature of the content (push-mode/pull-mode and progressive/interlaced display), the capability to reach the intended audience, and pricing.

AT ISSUE

Progressive Versus Interlaced, Round 2

The debate over progressive and interlaced display did not end with the FCC's approval of a statement in the ATSC agreement permitting both methods. A computer industry coalition beginning with Microsoft, Intel, and Compaq presented its vision for digital TV at the NAB conference in April 1997. This group, called the Digital TV Team, argues again that broadcasters should broadcast in progressive display. The joint press release is on the Compaq web site at:

http://www.compaq.com/newsroom/pr/pr070497i.html

(or search on "digital and tv" on the Compaq home page).

continues

The digital TV experience will be displayed on PCs, not television sets. According to the Digital TV Team's view, this does not negatively impact the broadcasters. Only the consumer electronics manufacturers who make interlaced monitors will be the odd men out.

Of the approved 18 ATSC options, the Digital TV Team endorses the following progressive display formats:

Vertical Lines	Horizontal Pixels	Aspect Ratio	Frame Rate per Second
720	1,280	16:9	24
480	704	16:9	24, 30, 60
480	704	4:3	24, 30, 60

In addition, the Team endorses two interlaced formats, to support only legacy television. All new broadcasts will be in the progressive formats.

The broadcasters continue to argue for high-definition interlaced displays because of its better pictures for most scenes, its lower bandwidth utilization, and the unavailability of progressive production equipment. They also maintain that the problem of poor graphics quality on interlaced displays can be resolved. The battle lines are drawn—Microsoft, Intel, and PC manufacturers advocate the progressive scanning scenario; the broadcasters, Japan, and much of cable TV broadcasters are staying with interlaced displays.

Will content creators who want to broadcast their product be able to bypass cable/TV broadcasters and rely on the Internet instead? Some observers believe that for the Internet to function as a broadcast medium, IP multicast will be challenged to scale up dramatically, particularly if many sites originate multicasts. Currently, the Internet Mbone can reach perhaps 50,000 viewers connected through 2,500 routers using IP multicast. The scaling to millions has yet to be demonstrated, but it might be possible if the required number of multicasts is relatively small.

Internet Content on Broadcast TV

Just as some broadcast content might bypass cable and TV in favor of Internet distribution, it might be possible for some Internet content to bypass the Internet and distribute over broadcast TV. Push-mode data, such as that provided by PointCast, can be distributed over the airwaves or by cable economically. Because PointCast does not use IP multicast, it consumes so much bandwidth that some corporations prohibit its use. Furthermore, most users of PointCast receive it over wide-area links, which operate at 1.5 Mbps or slower. Using broadcast SDTV, it would be possible to deliver up to 27 Mbps and offload the links subscribers have to their ISPs.

Many push-mode data services—such as stock tickers, weather services, and travel information services—can be distributed in this manner. This could have important implications for the development of datacasting, the popularity of SDTV, and the role of the Internet.

CONTENT SCENARIOS FOR ACCESS NETWORKS

The Access Network providers of integrated RBB systems must decide which type of content they want to convey and what transition scenarios to develop for their networks over time.

Full-Service Networking

A major issue is whether the Access Network provider will offer full-service networking. *Full-service networks (FSN)* are Access Networks that are intended to provide everything, including broadcast TV, Internet access, video on demand (VoD), and even voice telephony. The candidates are FTTC, vDSL, FTTH, cable, and LMDS. All are interactive—which is necessary for Internet services—and all have at least 51 Mb of bandwidth in the forward direction to carry multiple channels of broadcast TV. In the case of cable TV networks, providers intend to offer narrowband services and analog television as well.

Proponents argue that FSNs offer the lowest-cost alternative for subscribers to receive all services, certainly lower than subscribing independently to multiple Access Networks. They also argue that it would be easy to use for the subscriber, who would have a single interface to a variety of services. Finally, for the carrier, offering an FSN would tend to reduce customer churn, because a consumer would be more reluctant to part with a single service if that service came in a bundle. The control of customer churn is an important marketing requirement of any monthly service.

Of the Access Networks mentioned as FSN candidates, FTTC, vDSL, and FTTH can provide the least amount of bandwidth to the residence and thus are least suited to multichannel broadcast TV. Therefore, these Access Networks would be best suited for customers requiring VoD, high-speed Internet access, and narrowband services (such as telephony and metering) over the wired network. Users of these networks would have to obtain broadcast TV some other way (probably over-the-air), because HFC in combination with these other networks would be largely redundant.

Cable TV and LMDS are the networks particularly well suited to full-service networking because they are the only networks capable of two-way service and multichannel broadcast television.

Table 8–2 compares the capabilities of HFC and LMDS, with particular attention to bit rates.

Table 8–2 *HFC and LMDS Bit Rates*

	HFC	LMDS
Downstream bandwidth	600 MHz Assuming 54 MHz–654 MHz	850 MHZ Assuming 27.50 GHz–28.35 GHz
Downstream modulation scheme	QAM 64	QPSK

Table 8–2 *HFC and LMDS Bit Rates, Continued*

	HFC	LMDS
Downstream bit rate per second, net of FEC	2.7 Gb	1.4 Gb
Possible downstream utilization	200 digital TV channels +2.1 Gb for data and VoD	200 digital TV channels +800 Mb for data and VoD
Upstream bandwidth	6 MHz per 6 MHz of return channel	300 MHz
Upstream modulation scheme	QPSK or QAM 16	QPSK
Upstream bit rate per second, net of FEC	10 Mb	400 Mb
Number of homes sharing bandwidth	1,000 Density ranging from 500–2,000 homes per fiber node	1,600 Assuming 400 homes/square km with one transmitter serving 16 square km and 90° sectors
Average upstream bit rate per home passed	10 Kb	250 Kb
Average upstream bit rate per subscriber assuming 25 percent take rate	40 Kb	1.0 Mb

Table 8–3 compares bit-rate capabilities for FTTC and FTTH.

Table 8–3 *FTTC and FTTH Bit Rates*

	FTTC Profile A	FTTH (PON)
Downstream bit rate per second	51.8 Mbps	155 Mbps (OC-3)
Possible downstream utilization	5 digital TV channels +20 Mb for data	10 digital TV channels +100 Mb for data

Table 8–3 *FTTC and FTTH Bit Rates, Continued*

	FTTC Profile A	FTTH (PON)
Upstream bit rate per second	19.44 Mbps	155 Mbps (OC-3)
Number of homes sharing bandwidth	8–16	8–16
Average upstream bit rate per subscriber	1.5 Mbps	10 Mbps

Case Study: Time Warner Pegasus

A leading example of a system offering video services and inter-activity is the Time Warner Pegasus system. It is not strictly an FSN in that it does attempt to offer telephone service. Further, high-speed data services are offered by another Time Warner company called Roadrunner, which shares a common ATM backbone but uses the access network differently. Nonetheless Pegasus is an interesting case because of its plans to carry broadcast video, video on demand, and IP connectivity on a single network. A closer look at this particular network reveals some of the general questions and problems facing network implementors who are contemplating this most complete and integrated option of RBB methods.

After some experience with their preliminary FSN trial in Orlando, Florida, Time Warner will implement a new end-to-end, full-service network called Pegasus (see http://www.timewarner.com/rfp). Whereas the original Orlando trial used ATM end to end, Pegasus uses ATM in the Core Network, and MPEG and IP in the distribution network. At the set top, video is separated from data using frequency-division multiplexing or MPEG demultiplexing for in-band data.

The Orlando FSN proved the technical capability of HFC as a full-service network. Its costs were high because it was an innovative experiment. Pegasus is billed as a deployable form of the Orlando FSN.

The Pegasus system is designed to provide services for analog TV, digital TV, VoD, and two-way data service for control purposes mainly, but it also provides services for minor data applications through a single set-top box.

These services are provided through three types of channels. The *Forward Application Transport (FAT)* channel is a 6 MHz channel encoded using QAM 64 or QAM 256. VoD and broadcast TV are distributed on the FAT channels. The *Forward Data Channel (FDC)* is a 1 MHz channel encoded using QPSK to yield a T1 data channel in the forward direction. The *Reverse Data Channel (RDC)* is a QPSK-modulated 1 MHz T1 data channel in the reverse direction and is compatible with the DAVIC signaling specification. The FDC/RDC provide:

- Two-way real-time signaling for video on demand

- Delivering of set-top authorization messages

- Network management of the set-top

- IP connectivity for set-top applications—for example web browsing (in conjunction with the FAT channel for high-bandwidth services)

Figure 8–3 illustrates the major components of the Pegasus network. Analog broadcast is received at the head end and retransmitted on a FAT channel. Digital broadcast is received at the head end in MPEG packets with QPSK modulation. Satellite delivery generally is encoded in QPSK. Cable delivery is encoded using QAM 64. In addition, the encryption techniques used by satellite differ from those used by cable. The *Broadcast Cable Gateway (BCG)* remodulates and re-encrypts the satellite digital video into a form standardized on cable networks.

Figure 8–3
Pegasus full-service network.

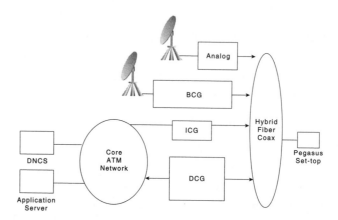

The *Interactive Cable Gateway (ICG)* is used in the early phase of Pegasus to enable the integration of video with data into the FAT channels. This is required, for example, to embed web content into a program or to enable electronic commerce if it needs video. Video and data are received from the Core Network over ATM at links of 155 Mbps. ICG is required to match the speeds of 155 Mbps to the FAT channel, which runs at 27 Mbps. Newer versions of Pegasus will require that media servers perform the integration of video and data onto a FAT channel by using specifications standardized by the DVB.

A *Data Channel Gateway (DCG)* provides a two-way real-time data communication path between the Pegasus set top and both the application servers (which perform such tasks as program selection) and the *Digital Network Control System (DNCS)*, which performs set-top authorization, set-top management, and purchase information collection. The DCG is essentially an IP router with QPSK interfaces.

A schematic of the Pegasus set top is shown in Figure 8–4. Beginning on the left side of the schematic, cable input is received from the drop cable through the QPSK receiver (FDC channel) and the RF tuner (FAT channel). An integrated MPEG and IP transport processor performs PID selection and forwards MPEG packets to the MPEG decoder and graphics processor, and forwards IP packets to the CPU bus.

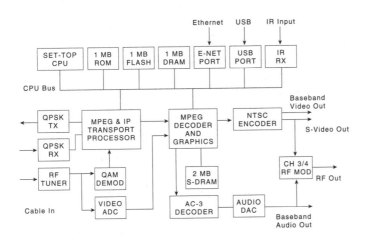

Figure 8–4

Pegasus set top.

Courtesy of Time Warner Cable (http://www.timewarner.com/)

For MPEG processing, a digital TV channel is demodulated in the QAM Demod and forwarded to the MPEG and IP transport processor. Audio PIDs are directed to the AC-3 decoder and then to an Audio Out socket. Video PIDs are decoded in the MPEG decoder and the graphics processor, then are fed to an NTSC encoder, and finally are sent to Video Out (either S-Video, baseband, or RF).

Analog video is received on a FAT channel and forwarded to the Video ADC, the analog-to-digital converter. The digital feed goes to the MPEG decoder, where it is processed as digital video.

IP packets are forwarded to the CPU bus and then to a set-top CPU. Outputs are available for Ethernet and *Universal Serial Bus (USB)*. USB is a new peripheral attachment standard that provides for 12 Mbps. A USB socket on Pegasus will enable standard peripherals, such as keyboards, joysticks, and printers to be connected to the Pegasus set top. (Further information on USB is available at `http://www.usb.org`.) Infra-red input (IR input) is received by the set-top box for television controls. Controls are sent in IP packets upstream through the QPSK transmitter (RDC channel). Finally, an Ethernet connector would be used to connect Pegasus to a PC or a game console.

Note that in this architecture no requirement exists for routing a Residential Gateway (RG). A network with this amount of bandwidth might not require downstream demultiplexing, because each piece of terminal equipment (TE) can be granted a forward data stream from the carrier premises. Consider the case of 500 residences sharing 2.7 Gb, each residence with 2.7 active users watching TV or browsing the web. In this case, each active user in the neighborhood can be granted exclusive use of 2.0 Mb with the downstream demultiplexing performed at the head end rather than in the residence. This moves intelligence out of the home and into the carrier premises, thereby reducing total system cost.

With the introduction of digital broadcast, more channels will originate in digital form, and the amount of analog bandwidth will drop over time. In this environment, cable networks can achieve their FSN promise.

Pegasus is expected to launch in late 1997. This successor to the original Time Warner FSN trial shows Time Warner continuing to press its lead in full-service networking.

Benefits of Full-Service Networking

FSNs tend to reduce complexity to the consumer (because of only one network to deal with) and reduce overall costs to consumers who want a wide variety of services. A single NIU, RG, and STU is amortized over a variety of services. There is also a single billing system and customer service infrastructure. All-in-one service motivates long-distance carriers to offer combined voice, Internet, paging, and voice service. Instead of dealing with a number of providers, the consumer pays one bill and possibly achieves certain discounts accordingly.

For the carrier, FSNs tend to create better account control and reduce customer churn. A customer cannot simply drop one service; he must drop a bundle of services in the event he wants to change carriers. FSNs also provide a better marketing information database. By offering more services to the subscriber, more is known about the consumer, and therefore it is possible to target the market more aggressively.

Challenges to Full-Service Networking

Although Pegasus and the overall work of Time Warner show a vision and commitment to RBB, significant questions remain about FSNs in general.

Are There Enough Residential PCs?

FSNs will provide both TV and computer service. If the density of PCs is too sparse, then the cost of passing homes is a big hurdle to overcome. With roughly 70 million households worldwide with PCs and about 35 percent of homes in the United States with a PC, the question is whether this distribution is too thinly distributed for FSNs. PCs are distributed around the country in many high-income neighborhoods nationwide. It remains to be seen if the enclaves of PC penetration are large enough to support FSNs in any single area.

If the residential neighborhood has low PC penetration, interactivity can be provided through the television set. But the business question remains whether consumers who are not interested enough or cannot afford to buy a computer maintain long-term connectivity to an interactive service.

Place Your Bets Up Front

Many carriers fear that moving to a full-service network is an all-or-nothing proposition—place your billion-dollar bets up front. It will take two to three years to see what competition emerges and five to ten years to get an accurate picture of financial results.

It would be financially less risky if the carrier could add services and bandwidth capacity incrementally. But FSNs must develop capabilities for multiple services rather than developing a single capability, such as voice only or data only. In addition, those combined services depend on having all the required bandwidth at the outset. For both services and bandwidth, a significant up-front investment is required.

Incomplete information on customer preferences, pricing sensitivity, competitive landscape, scaling properties of the technology, and the regulatory environment make FSN seem like a long-odds gamble to many companies. As an indicator of what other potential carriers can expect, the Time Warner FSN in Orlando was not an entirely reassuring experiment. Only familiar services such as VoD proved popular. Unfamiliar services such as interactive shopping and news-on-demand did not fare well. It is unclear whether the market drivers mentioned in Chapter 1, "Market Drivers," will in fact be popular enough in the early years of RBB rollout to keep the carriers' interest.

Biting Off Too Much

Telephony, video, and data delivery systems individually are large systems that have been subject to optimization for years by lots of talented people. When optimizing large systems, it is not necessarily true that one obtains a globally optimal result by optimizing everything at once. Sometimes better results are achieved by segmenting the problem, optimizing each subproblem independently, and then adding the results together. This is especially true when each subsystem is substantially different. Instead of trying to optimize voice, video, and data simultaneously in a single system, it might be more cost-effective to optimize each system independently and simply provide three networks.

Transition Scenarios

As they contemplate becoming part of a larger RBB network, existing Access Networks must plan how they will evolve. Local exchange carriers (LECs) can transition to wireless networks for video and xDSL networks for data without offering full-service networks. This is in response to their concern about time to market and their inability to install sufficient cabling to support HFC or FTTC in the near future. As markets for video and xDSL develop, the telephone companies can move onto LMDS or FTTC. Ultimately, FTTH is viewed as the end game for carriers, although they might stay with wireless as those technologies improve their robustness.

Cable operators basically must evolve to the HFC model. As fiber-node clusters reduce in size, they eventually look a lot like FTTC networks. But cable operators' experience with multichannel delivery likely will preclude delivering FTTC service as they remain in broadcast mode in the downstream direction.

Figure 8–5 outlines likely evolution paths for current technologies.

Figure 8–5
Transition scenarios.

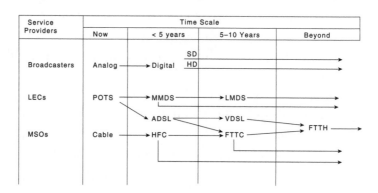

VIEWPOINT

RBB as Defense

With the exception of FSNs, the RBB business scenarios of most carriers can be characterized as defensive. For the most part, service providers will not deploy new services in areas where they are not challenged. Pacific Bell, for example, is rolling out ADSL and HFC services in areas where cable is forging ahead with cable modems. But where cable is relatively inactive, the LECs have initiated little discussion of ADSL. The LECs would rather invest in long-distance service, a service that involves relatively little incremental cost, has similarities to current services, and promises a known and large-revenue potential.

Cable must defend itself from DBS in the broadcast TV space. Some cable operators have adopted the defensive strategy of introducing their own DBS service.

A mainly defensive approach tends to slow the pace of RBB rollout. If one major industry group slows down, others tend to slow down as well, waiting for applications to develop that force them to progress. Internet access is an application that developed ahead of the network to provide it. The same kind of process might be necessary to push the carriers to move beyond just defensive deployment strategies to more aggressive market expansion.

PREDICTORS OF RBB SUCCESS

As they evolve into integrated RBB networks, existing access and content providers are trying to gauge their own chances of success against the competition. In nearly every major urban area of the United States, and probably other parts of the world as well, there will be three to six providers of high-speed residential services. The following list details possible contenders:

- DBS service offered by two to three national providers

- Either MMDS or LMDS service from out-of-region telephone companies, interexchange carriers, or content providers seeking to bypass the telco/cable providers

- HFC service from cable operators

- ADSL service from the LECs, an independent Internet service provider, or competitive local exchange carriers (CLEC)

- Later, FTTx services from the LECs

In addition, narrowband service will be available from the electric utility and several telephone companies. Telephone service can come from LECs, interexchange carriers, cellular carriers, digital PCs, and CLECs. This adds up to a dozen competitors for the combined residential voice, video, and data market in a specific area. Given this competitive landscape, one must wonder which Access Networks and content providers will succeed in the short term and over the long term. The following questions and answers provide some tools for predicting success or failure.

Question 1: *What Networks Are in the Weakest and Strongest Positions as Technology Changes?*

Technological change and its twin, technological obsolescence, tend to reduce the usable life of equipment and thereby increase expense. This fact would tend to favor models in which cost is pushed to the consumer and models in which vendors can withstand heavy depreciation expenses. It also puts a premium on customer service—and, to some extent, remote management—for Access Networks that don't have the latest technology and can't upgrade.

Technological innovation as a market driver is less important in the consumer market than in the business market. Except for a few early adopters, consumers aren't inclined to purchase new products and services for the sake of being on the cutting edge. Content and access providers must take care not to use technology as the centerpiece of their marketing.

Question 2: *Who Can Withstand Churn?*

Churn (or customer turnover to other vendors of similar services) has long been a marketing concern of service providers, particularly long-distance telephone and ISPs. There will be less churn for RBB services in which the consumer must purchase equipment up-front to obtain service. This is the case for DBS satellite. Cable operators also are promoting a retail distribution model for set tops, which would have the same effect.

But imposing an up-front purchase by the consumer limits initial sales. Some analysts believe that DBS cannot exceed market share of 15–25 percent of households without reducing dish prices to below $100.

Because limits exist as to how much up-front investment carriers can impose on consumers without jeopardizing sales, carriers will have to find other ways to cope with churn as well. Specifically, the Access Networks must minimize the sales and administrative costs of signing up new subscribers. Efficient installation processes also will be critical. These are advantages of incumbent service providers with historically strong operational support systems, such as phone companies.

Question 3: *Who Has the Most Endurance?*

Patience is a key success factor when dealing with big unknowns and big investment. When observers and investors ask, "Which Access Network wins?" the answer is based largely on how much pain the service provider can withstand and for how long. This cannot be predicted perfectly; one can consider only relative strengths and weaknesses. The LECs have money; the MSOs have content. The broadcasters have both but no interactive future. The ISPs have content, but little of it is amenable to the advertising model. In any case, the ISPs don't own the infrastructure. Each of these providers has strengths and weaknesses, but what seems more important is how long each will persist.

Question 4: *Who Will Optimally Match Content to Distribution?*

Ultimately, when dealing with consumer markets, the win comes by matching content to the appropriate distribution method.

Broadcast TV was the technological fit for passive mass audience because of its one-way distribution of video and sound. From its earliest days to the present—from *I Love Lucy* to *The Simpsons*, from the first televised Olympics to the most recent NBA finals—television's programmers have sought to tap into broad, shared entertainment preferences. Broadcast TV was and is also an ideal advertising delivery mechanism, a fact that has subsidized it and ensured its continued existence.

The Internet is the technological fit for the active, two-way, point-to-point dissemination of information. That information is useful for research and business. The Internet could be a platform for entertainment as well, especially entertainment tailored to individual tastes and interactive participation. It, too, is an example of a technology that evolved as an optimal combination of content and distribution.

RBB will take whatever form is needed to deliver the critical market drivers, the new content and applications that will drive its development.

High-definition content would tend to favor higher-bandwidth networks, particularly cable and LMDS. Chat-room activity combined with broadcast TV favors networks that can provide integrated voice services. These likely would be ATM-based networks that can run over cable, wired networks, or wireless networks.

Virtual channels don't require the speed of cable and LMDS, so xDSL could be a cost-effective method of delivering this application. If Internet access alone is preferred and viewers obtain their TV over-the-air, networks with a strong return-path story have an advantage.

It is not clear at this point what the dominant market drivers will be. For that matter, new applications and content are certain to emerge. The point is that there is no one best Access Network for all services. Some technologies fit certain content better than others.

The success of RBB depends on finding the market drivers. The technology will follow.

SUMMARY

As complex as the technical, regulatory, and business questions are for each of the networks described in Chapters 3–7, additional challenges—including software issues—arise when it comes to integrating content and distribution networks. In the end, access providers must decide what kind of content to carry, given their current and evolving capabilities. Matching the content—the significant market drivers—to the appropriate RBB delivery system is the key predictor of which RBB networks will thrive.

REFERENCES

Article

Laubach, Mark. "Serving Up Quality of Service." *CED Magazine*, April 1997.

Documents

Digital Audio Visual Council (1996), DAVIC 1.1 Specification Part 8: Lower Layer Protocols and Physical Interfaces (draft as of September, 1996).

ITU-T Recommendation J.83 Annex B (1995), Digital Multi-Programme Systems for Television Sound and Data Services for Cable Distribution.

ISO/IEC 13818-6 (1995), Information Technology—Generic Coding of Moving Pictures and Associated Audio: Extension for DSM-CC.

ISO/IEC 13818-2 (1994), Information Technology—Generic Coding of Moving Pictures and Associated Audio: Video.

SCTE DVS-031, Society of Cable Television Engineers, Digital Video Transmission Standard for Cable Television—Digital Video Standards, October 15, 1996.

Time Warner Cable—Pegasus Program Request For Proposal and Functional Requirements, Time Warner Cable—Engineering & Technology, March 6, 1996.

Conference Proceedings

Adams, Michael. *Real Time MPEG Asset Delivery over ATM*, Proceedings National Cable Television Association—Cable 1995 Conference, Dallas, TX. May 1995, pages 315–326.

Brown, Ralph W. and John Callahan. *Software Architecture for Broadband CATV Interactive Systems*, Proceedings National Cable Television Association—Cable 1995 Conference, Dallas, TX. May 1995, pages 270–278.

Williamson, Louis D. *Conditional Access and Security Considerations of Transactional Broadcast Digital Systems*, Proceedings National Cable Television Association—Cable 1995 Conference, Dallas, TX. May 1995, pages 207–212.

Internet Resources

http://www.atsc.org
(System Information for Digital Television, ATSC Standard, Document A/56, January 3, 1996.)

http://www.powertv.com
(Information on the PowerTV™ operating system.)

http://www.microsoft.com/supercomm
(Microsoft's case for ATM end to end.)

RBB-Related Companies and Organizations

Space limitations do not permit a comprehensive listing of all the companies and organizatons that are playing a role in the development of residential broadband networks. The information in this appendix is intended as a representative cross-section of some of the most active players, including all those that are cited elsewhere in this book.

Table A–1 *Equipment Providers*

Company	Company	URLs	Stock Ticker
3Com	Cable modems, ATM switches	www.3com.com	coms
@Home	Systems integrator to provide data services over HFC networks	www.home.net	athm
Adaptec	Firewire components	www.adaptec.com	adpt
Alcatel	Systems for xDSL, FTTC, and FTTH, ATM switches	www.alcatel.com	ala
Amati Communications Corporation	xDSL systems	www.amati.com	amtx
Analog Devices	xDSL systems	www.analog.com	
Apple	Firewire components	www.apple.com	aapl
Aware, Inc.	xDSL systems	www.aware.com	awre
Bay Networks	Cable modems, ATM switches	www.baynetworks.com	bay
Bosch Telecom	LMDS systems, acquired from Texas Instruments	www.bosch-telecom-leipzig.mda.de	
Broadcom	QAM modulators and demodulators	www.broadcom.com	brcm
Cisco Systems	Cable modems, xDSL systems, residential gateways, ATM switches, Internetworking software	www.cisco.com	csco
Com21	Cable modems	www.com21.com	
Comcast	Cable MSO, partly owned by Microsoft	www.comcast.com	cmcsk
Cox Communications	Cable MSO	www.cox.com	cox

Table A–1 *Equipment Providers, Continued*

Company	Company	URLs	Stock Ticker
DiviCom	MPEG encoders and decoders	www.divi.com	
General Instrument	Cable modems, cable system components, MMDS components, FTTC systems in its Next Leel subsidiary, digital set tops	www.gi.com	gic
Hewlett-Packard	LMDS systems, test equipment for access networks	www.hp.com	hwp
Lucent Technologies	DLC systems, ATM switches	www.lucent.com	lu
MediaOne	Cable MSO, formerly Continental Cable	www.mediaone.com	
Microware	OS for digital set tops, called David	www.microware.com	
Mitsubishi Electric	Digital TV components, Firewire components, residential gateways	www.melco.co.jp	
Motorola	Cable modems, xDSL systems, LEO systems, equity owner in Iridium	www.motorola.com	mot
Netspeed	ADSL systems	www.netspeed.com	
Nortel (Northern Telecom)	DLC systems, ATM switches	www.nortel.com	nt
Orckit Communications Ltd.	xDSL systems, VDSL components	www.orckit.com	orctf
PairGain	xDSL systems	www.pairgain.com	pair
Paradyne	xDSL systems, leaders in CAP	www.paradyne.com	

Table A–1   *Equipment Providers, Continued*

Company	Company	URLs	Stock Ticker
PowerTV	OS for digital set tops	www.powertv.com	
Scientific Atlanta	Cable modems, cable system components, digital set tops	www.scientific-atlanta.com	sfa
SGS-Thomson Microelectronics	Microelectronics, holding company for TCE and TSI, Firewire components	www.st.com	stm
StarSight	Electronic program guides	www.starsight.com	sght
TCI (Tele-Communications, Inc.)	United States' largest cable MSO	www.tci.com	tcoma
Terayon Corporation	Cable modems and ASICs, specializing in spread spectrum	www.terayon.com	
Texas Instruments	xDSL, DSPs, Firewire components	www.ti.com	txn
Thomson Sun Interactive	OS for digital set tops, called OpenTV	www.opentv.com	
Westell	xDSL systems	www.westell.com	wstl

Table A–2   *Professional Organizations and Regulatory Agencies*

Organization/Agency	Description	URLs
ADSL Forum	Developing standards for ADSL	www.adsl.com
Advisory Committee on Advanced Television Services (ACATS)	Advisory committee to the FCC to advise the Commission on HDTV. More information available on the FCC's web site	
The ATM Forum	Developing standards for residential networking based on ATM	www.atmforum.com

Table A–2 *Professional Organizations and Regulatory Agencies, Continued*

Organization/Agency	Description	URLs
Computer Industry Coalition on Advanced Television Service (CICATS)	Apple, Compaq, Cray, Dell, Hewlett-Packard, Intel, Microsoft, Novell, Oracle, Silicon Graphics, and Tandem weigh in on the future of TV	
Digital Audio Visual Council, or DAVIC	Developing standards for access networks, home network	www.davic.org
Digital Video Broadcasting Project (DVB)	European organizations specifying the future of television	www.dvb.org/dvb_index.html
European Telecommunications Standards Institute (ETSI)	Telecommunications standardization body	www.etsi.fr/dir/about/about
Federal Communications Commission	The rules of the road for wireless and cable in the United States	www.fcc.gov
IEEE 802.14	Developing a standard for cable data modems	www.walkingdog.com
International Electrotechnical Commission (IEC)	Electronics standardization in Europe	www.iec.ch
MCNS Partners Ltd.	Developing a purchasing specification, not a standard, for cable data modems	www.cablemodem.com
National Association of Broadcasters (NAB)	Trade association and lobbying arm for U.S. radio and television broadcasters	www.nab.org
Video Electronics Standards Association	Specifying standards for digital set tops	www.vesa.org

Table A–3 *Service Providers*

Company	Service	URLs	Stock Ticker
Ameritech	Local exchange carrier, cable operator	www.ameritech.com	ait
AT&T	Interexchange carrier	www.att.com	t
Bell Atlantic	Local exchange carrier	www.bellatlantic.com	bel
Bell South	Local exchange carrier	www.bellsouth.com	bls
BTnet	UK's dominant service provider, equity interests in DBS FTTx trials	www.bt.net	
Canal+	Cable system in France	www.cplus.fr	
CellularVision	LMDS service provider	www.cellularvision.com	cvus
DF1	DBS service provider in Germany	www.dfl.de	
DirecTV	DBS service provider	www.directv.com	
Echostar Communications Corporation	DBS service provider	www.echostar.com	dish
Globalstar	Low earth orbit service provider	www.globalstar.com	gstrf
GTE	Local exchange carrier	www.gte.com	gte
Iridium	Low earth orbit service provider	www.iridium.com	
MCI	Interexchange carrier	www.mci.com	mcic
News Corporation	Rupert Murdoch's vertically integrated, global satellite, broadcasting, publishing Fox network and *New York Post* in the United States	www.newscorp.com	nws

Table A–3   *Service Providers, Continued*

Company	Service	URLs	Stock Ticker
NTT (Nippon Telegraph and Telephone Corporation)	Japan's dominant service provider, FTTx trials	www.ntt.co.jp	
NYNEX	Local exchange carrier, FTTx trials	www.nynex.com	nyn
Pacific Bell	LEC, MMDS service provider, Local exchange carrier	www.parbell.com	pac
PointCast	A provider of push-mode data on the Internet, which does not use multicast	www.pointcast.com	
Primestar	DBS service provider	www.primestar.com	
Rogers	Cable system in Canada	www.wave.ca/Homenw.html	
SBC Communications Inc.	Local exchange carrier	www.sbc.com	sbc
Sprint	Interexchange carrier	www.sprint.com/index.html	fon
Time Warner	U.S. number 2 cable MSO; content owned includes Warner Brothers, Castle Rock, Turner, TNT Broadcasting, Hanna Barbera, *Time*, *Life*, *Fortune*, *People*, *Sports Illustrated*, many others	www.pathfinder.com www.timewarner.com	twx
Teledesic Corporation	Low earth orbit service provider	www.teledesic.com	
US West	Local exchange carrier, cable operator	www.uswest.com	usw

Table A–3   *Service Providers, Continued*

Company	Service	URLs	Stock Ticker
US West Media Group	Cable system, mostly consisting of MediaOne holdings, TeleWest (UK) cable, Time Warner Cable, wireless and web services	www.uswest.com/media/index.html	umg
Vidéotron ltée	Cable system in Canada	www.videotron.ca	

Index